I0500229

UNIVERSAL THEORY

A MODEL FOR THE THEORY OF EVERYTHING

MOHSEN KERMANSHAHI

Universal Publishers
Boca Raton, Florida

Universal Theory:
A Model for the Theory of Everything

Copyright © 2007 Mohsen Kermanshahi
All rights reserved.

Universal Publishers
Boca Raton, Florida
2007 • USA

ISBN: 1-58112- 943-2
13-ISBN: 978-1-58112-943-4

www.universal-publishers.com

To my dear wife Firoozeh

And precious children

Sanaz, Sarvenaz and Sam

Table of Contents

5

Preface

"Where did I come from? Why did I come?
Will you tell me where my final destination will be?"
Jalaluddin Rumi 13th Century Persian poet

Who are we? Where are we? Why are we here?
Like everyone else entering the prime of their life, I have established a lifetime of success and failure, happiness and sorrow, moments of turmoil and times of order and routine. And although I had seldom a chance to pause and think about the meaning of life or the nature of the things around me, these issues have always preoccupied my thoughts. When the time came, this new era of my life brought the desire to find the answers to these questions that have preoccupied many before me, and will imminently occupy the minds of many to come. But then, where would I find the answers for these fundamental questions? Many people follow a spiritual path, but I decided to look for the answers in the domain of science.

We are in the dawn of the twenty first century where science has shown to have evolved a great deal, at least since I was in school about forty years ago. Puzzles like those we have encountered in Quantum Mechanics, new findings in cosmology, and research into the conscious mind take us to mysterious places and present domains previously unimaginable. There is a lot to learn and understand for someone like me who is in the pursuit of uncovering answers to fundamental questions. Have I been successful in finding answers? Not entirely, though factually, we may never know the whole truth, however there is no doubt that the more knowledge we attain, the deeper our insights into the true reality that awaits us. I can however, say with conviction that I have come a long way.

During the course of multidisciplinary studies, this concept formed in my mind a little at a time. This model was constructed through the over-viewing the different branches of human knowledge. What motivated me to pursue this idea further was that this developing vision in my mind could offer potential explanations for many of presently unresolved issues in science.

The more I studied, the more I felt that this concept could be a viable model. I strongly believe that the concept presented here can help us in attaining a more profound understanding of reality.

Over the course of the past five years, I have been developing this model step by step I have been especially encouraged by positive comments written by readers who have been visiting the web site where the concept has been posted, www.universaltheory.org. Some chapters of the book were also posted in a very wonderful website called TOE quest. This website -www.toequest.com- has been developed by Robert Armstrong and is devoted to discovering the Theory of Everything. This is a theory in which physicists from Einstein onwards have been exploring and is supposed to remove discrepancies between the two main branches of physics, Quantum Mechanics and Astrophysics. I am proud to have been chosen as the most favorite author on the TOE quest website, and also proud that my concept is consistent, coherent, fine tuned, and now at the stage of completion. This model offers an alternative for physical reality as we know it. If proved to be sound it could contribute greatly to human philosophy and will give insight to the reader for a more profound understanding of reality.

I have to thank Nahid Sahel Gozin who have written up the innovative mathematics for this model. Nahid is a physicist who has helped keep me from drifting astray during the course of writing this book. Many of the artistic drawings in this book were also prepared by Nahid. Without Nahid's tremendous support, this concept could not have been developed to this extent.

I would like to thank Eugenia Tang who has edited the pages and made this book readable and comprehensible. Above all, thanks to all the members of the scientific community, past and present, who have opened our eyes to new horizons and have made the universe a more understandable and malleable concept for the human race.

Introduction

"Imagination is entrusted with the responsibility to explore. Its mission is not to abjure reality but, rather, to magnify our intercourse with it. Karl Popper and Gaston Bachelard, among the others, have emphasized the historic importance of speculation in forming hypothesis and establishing scientific truth."

Etienne Klein

The pursuit to understand reality is a common quest. The human race has tried to understand the world and beyond since the beginning of its history. We have envisioned different scenarios over the years past. Have we come far enough? Certainly, we can be proud of our achievements. But amazingly, new and very interesting windows have been opened in front of our eyes. The new frontiers are very unfamiliar and not comprehensible to our traditional understanding.

Quantum super-position of states, the Heisenberg Uncertainty Principle, quantum entanglement, and other characteristics of the sub-atomic arena have introduced an in-deterministic interpretation of reality. This has left objective physicists uneasy. Numerous attempts, such as Bell's Theorem, have failed to extend the rules and logic of familiar classical physics to the subatomic domain.

New puzzles also come about in Astrophysics. Black holes, dark matter, non-zero cosmological constant and dark energy, etc. has complicated the scheme even further.

The scientific method of psychological assessment has also unveiled a new understanding of the mind and consciousness. Many experiments support notions like Holonomic Brain theory (non-locality of consciousness) and transpersonal psychology (a universal Psyche).

Certainly, solving the above puzzles is the un-chanted road ahead. How can we go about doing so. The unexplained has left the science as an open system. Efforts have been directed at finding explanations for such strange phenomena in the last century with traditional scientific method of thoughts but to no avail. At this

9

point it seems that we have to alter our vision and logic in order to interpret new findings. A new vision will help us interpret the unexplained, so that we can return to a scientifically closed system. To this end, I believe, we have to extend the observable arena and introduce definitions for zeros and infinities. Unfortunately so far mainstream physics and the dominant theoretical physical theories have chosen to avoid these two fundamental elements.

This model is in line with Einstein's view, which presumes the presence of reality out there beyond our consciousness and incorporates Neil Bohr's view because it has a mind component.

Aims

In this book, I am presenting a concept, which can reintroduce objectivity as a tool to explore reality. I believe this model has the potential to bring us back to a deterministic world.

Together we will revisit the origin of the universe with a new vision. We will explore whether this new perspective is capable of offering answers for different astrophysical, quantum mechanical and psychological paradoxes.

The effort to locate an explanation for the unexplained falls into the long sought after the Theory of Everything, which aims to solve the inconsistencies between Einstein's General Relativity and quantum mechanics (the study of subatomic particles). Despite endless efforts made by great physicists, Einstein's quest for a Theory of Everything still remains obscure. I describe one of the problems at this time as an example;

In General Relativity, space is presented has a smooth curve produced by the existing masses in it. This is in line with our everyday experience. As we go to smaller scales the smoothness continues till we reach ultra small or Planck scale ($1.6*10^{-33}$ cm). In the vicinity of this scale the fabric of space starts to get very rough and cranky. Gravity cannot explain the violence, which exists in fabric of space in ultra small scale. Interestingly, the violence exists even without an actual particle being present.

In Planck scale quantum mechanics prevail. The roughness of the fabric of space in ultra short scale is explained by quantum mechanics. However, the explanation about the shape of the space is not compatible in these two theories. The Theory of Everything is supposed to remove discrepancies and enclose the concepts of quantum mechanics and general relativity in one grand theory.

Ideas presented in this book aim at being a model for the theory of everything.

Avoidance

General Relativity precisely describes the motion of stars and galaxies as well as everyday events that we encounter in our lives. Quantum Mechanics on the other hand, accurately predicts and describes subatomic particle reactions and the world that exists in small scale. Ideally, they would have to work hand in hand to account for the entire universe as a whole. Unfortunately, we have not been able to find a feasible relationship between these two major domains of physics. In fact, they are currently considered incompatible. If we use the mathematics of general relativity and quantum mechanics together, the answer is usually *infinity*. Since infinity cannot exist in our present model of the universe, these calculations are discarded and incompatibility is declared.

It seems to me that we have consciously decided to avoid paying heed to road signs that are erected in every step of the way. This is all done by clinging to objectivity (Tangibility). The problem is that we are using traditional judgment and logic to define this objectivity. Existing theories constantly circumvent singularities and infinities that regularly present themselves in experiments and calculations.

During 15th Century mathematicians frequently found that positive numbers could not explain all of the functions that existed in the mathematical field. They realized that they had to expand the field to include a new domain. This domain would be different than the one that they had been acquainted with, so they extended the field to accommodate negative numbers. By doing so, many unexplained calculations could then be accounted for and subsequently understood.

Then once again, when faced with the square root of negative numbers ($\sqrt{-n}$), mathematicians realized that their understanding of math is not complete. So they added yet another arena. They opened themselves to the concept of the so-called imaginary numbers, although the concept was still mysterious and obscure at the time.

In the field of Physics, we have been unwilling to explore and adapt to the unknown arena. We have built many ideas and models but all of them are within a kind of space-time field. Despite the fact that known physics tells us that space, time and matter are not absolute and that most paradoxes arise when we get close to the boundaries of space-time, we are still afraid of going over the cliff.

Subconsciously we aim to construct theories that avoid delving into a field beyond that of the familiar, space-time.

The majority of attempts to find explanations for the unexplained, such as the super string and M theories, loop quantum gravity, etc. have been constructed within framework of this "known" arena.

Since our objective space-time knowledge deals with computable objects, the physical meaning of zero is not defined in our tangible world. In addition, because quantity in space-time is numbered and ultimately finite, we can not define infinities either. Thus in theoretical physics, we regularly eliminate and ignore them and as we call it (normalize them) in our equations.

A sure signal of direction is the fact that calculus has been the mathematics of choice to explain the fundamentals of theoretical physics. Differential calculus, derivatives, tangents, limits of sequence, zeno problems, and light cones all point to"zero". Instead of avoiding it we had better open our eyes and take a closer look at point zero and the fundamental role it plays in our physical world.

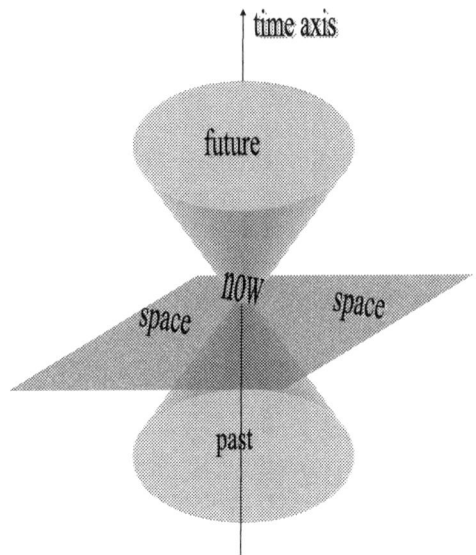

Light Cone

Many mechanisms have been adopted to bypass zeros and infinities. Even Schrodinger's equation, which helps us to understand and calculate quantum mechanical functions, has been formulated to help cliff settlers remain on their safe spot (familiar space-time) and continue their calculations in a safe and familiar environment. Theoretical physicists call these run away mechanisms "normalization". We hesitate to go over the edge of the cliff because we believe that, we are not capable of examination or measuring or even comprehend singularities or infinities. But read on and see if this kind of reasoning is in fact realistic.

All this is happening despite the fact that we have the appropriate tools to explore zeros and the non-computable. Imaginary numbers were introduced in the 16th century. The nature of complex numbers (a combination of real and imaginary numbers) was gradually constructed and today we have reached a point where we have a much clearer understanding of this system. As a matter of fact complex numbers are now a fundamental part of mathematics and physics at a deeper level. The complex number system can guide us to define zero and infinities in the physical world. That is if we open our scope and imagine a

13

physics that surpasses and goes beyond our space–time. The model presented in this book is based on an enclosed space-time.

Gordon Kane from university of Michigan in Ann Arbor believes:

> "As we go to smaller distances or higher energies, we expect each effective theory to need re-normalization this is not a problem or an unexpected failure of the theory. However, the primary theory (*Theory of everything*) had better not need such inputs or re-normalization." [50]

It is understandable if a midway theory, which is not able to go deep enough, requires re-normalization in order to present its mandate. However, as Gordon Kane states, the ultimate theory cannot ignore any portion of evidence and would have to explain every aspect of reality. Regrettably, he further argues: "It has to be a finite theory (One that never gives an infinite prediction for a physical quantity)."

If mathematics is the infrastructure and blue print for physical theories and zero and infinity are indispensable portions of mathematics, then we have to open our mind to the possibility that the ultimate theory may not be a finite theory and therefore may very well contain zeros and infinities.

Through the act of re-normalization, we have left physics as an open system. An open system by nature cannot contain and conclude the whole function. This kind of approach has left a good portion of reality out of our reach and understanding.

Not only have we not paid heed to the clearly visible road signs, but we have also attempted to cover them up. Theories like Super Gravity, Super Symmetry, Super String and the like, are introduced primarily to cancel out zeros and infinities. Maybe this is the main problem; it is dragging the energy and capacities of the mainstream physics community off track.

At the same time, a theory that can only solve the conflicts between two major components of today's physics, namely General Relativity and Quantum Mechanics, cannot be considered as The Theory of Everything. To be able to claim universality, such a theory must also explain the emergence of life and the mystery of mind. The old dilemma between consciousness being a separate entity or the product of the brain has reached new horizons.

Next I will start this venture by offering a fundamental theory utilizing intuitions derived from space-time knowledge. With the belief that cause and effect are a part of one system and by looking beyond the known realm, we should be able to collect evidence which guides us to a deeper understanding of reality. I intend to show that the space-time universe, the quantum world, and the mind are the tripod of reality. In order to achieve this deeper understanding of reality, we must study the above topics. I believe this to be the main challenge put forth for the human race in 21st century and beyond.

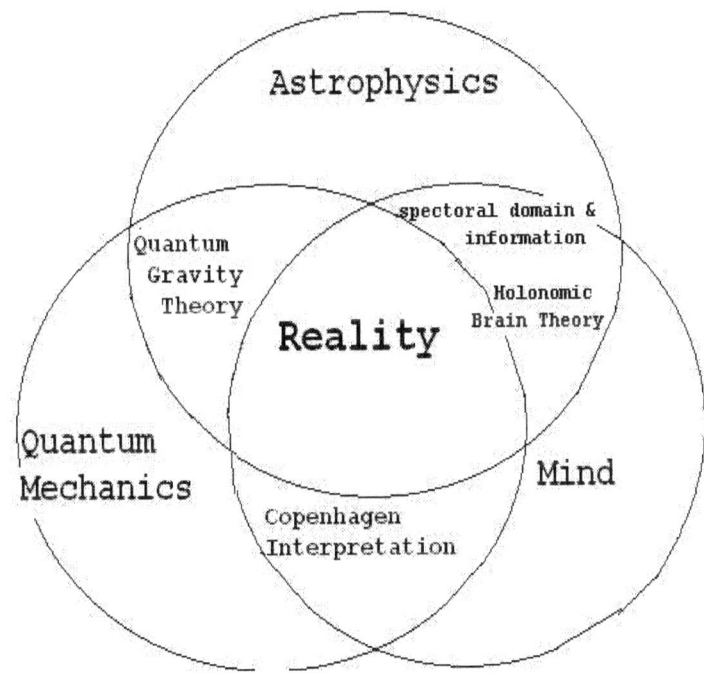

Reality Tripod

The above fields (even with our present knowledge) are derived, joined and united in a kind of singularity; a singularity with a definition different from that which is traditionally accepted. Singularity by the traditional definition is the ultra dense zero size point which is the initial nucleolus of our universe and the origin of the Big Bang.

Roger Penrose states:

"This new theory (theory of everything) will not just be a modification of quantum mechanics, but something as different from standard Quantum Mechanics as General Relativity is different from Newtonian gravity. It would have to be something, which has a completely different conceptual framework. In this picture, quantum non-locality would be built in the theory"[5]

Although we will make use of mathematical concept as often as we can, do not expect that this model be a traditional physical model, based on and derived primarily from a Lagrangian mathematical structure. One main reason for this is that the Platonic mathematical world does not match the physical world in its entirety. Roger Penrose looks at the Platonic Mathematical world as a perfect and separate entity. He believes our physical world is utilizing just a portion of that perfect world. [56]

An ideal sphere found in geometry

Maybe we can relate the difference between these two separate worlds to the fact that there are many more factors shaping the physical world as compared to the mathematical world. For example let us compare a geometrical sphere to a similar physical body like a planet.

An actual spherical object in space-time

The only determining factor of a sphere in geometry is its radius. But many elements are affecting the shape of a physical object like a planet. The gravity gives the mass of a planet a spherical shape whereas its spinning tends to somewhat flatten the poles of our otherwise perfect sphere. In addition, the mass constituents and grain in conjunction with geological and atmospheric activities change the interiors and surface texture of the planet. So a planet ends up being far from a perfect sphere. The planet is also alive. I mean like any other real object it has an imaginary (i) dimension. So in the physical world, not only are there many factors at work but according to the fundamental concept of complex numbers, objects have an imaginary dimension as well. Maybe that is why the physical arena is different from the precise and idealistic platonic mathematical world.

Therefore, I conclude that a less complicated mathematical construct cannot precisely represent the multi-factorial and complex physical domain. For all practical purposes, one can assume that the valid portion of mathematics is the portion that is consistent with the structure of the physical world. But there is a fine line in here. Often mathematical propositions guided the physicist to new discoveries. At the same time one should be cautious of capitalizing on complicated mathematical constructs that pulling us farther away from tangible physical world. Many times this kind of mathematics can be non-physical. May be this is what is happening to the leading candidate for the Theory of everything, the Super String Theory. Many of its components, which are based on mathematical propositions, are not observed.

Occasional use of math equations in this book should not be a deterrent to readers who are not as proficient in mathematics. At the end of each mathematical proposition there is an assertion written in plain English so the reader can follow through the discussion with ease.

I would like to clarify that this model is not a reductionists attempt to explain reality from the bottom up. And although I utilize experiments and evidence to substantiate the principles of this theory, at the present it is not a theory of physics, because the Lagrangian has not yet been written for it. However, it is consistent and coherent. I have tried to adhere with scientific methods in my study. This is a new speculation from a new axiom and I hope it proves itself helpful in explaining the unexplained.

The following model attempts to answer a wide range of mysteries, which are facing us today. Hereinafter I call it the Universal Model.

While we have tried not to stray from current knowledge, some of the ideas presented may not be completely agreed upon by mainstream scientists. Although the introduction of new ideas is the way for science to evolve the reader is cautioned to use his or her own judgment before adopting any of the presented concepts.

Complex Numbers

Co-written with Nahid Sahel Gozin

Contemporary physics has been developed over the centuries and has been quite effective as a base for science and technology. But many puzzles and paradoxes exist in quantum mechanics (the sub-atomic arena) and astrophysics which cannot be solved in the context of the current physics domain. Consciousness also remains a mystery. In fact, the findings of the past century defy the principles of contemporary physics. New revelations have painted a more detailed universe and resent an alternative reality that we cannot explain with the traditional axiom. An alternative physics is needed to explain the newfound reality. That is why new fundamental theories are being introduced to explain the unexplained.

In this book, I will present an alternative physical model for the universe and offer explanations for existing paradoxes based on this new concept. In this model, the space-time universe is embedded in a newly defined singularity. In order to better follow this model, familiarity with the concept of complex numbers is helpful. I will try to explain the concept in layman's terms for the reader who is not keen in mathematics. Alternatively, the reader may choose to skip the math equations and just look over the assertions made on their basis. Doing so will not prevent comprehension of the concept.

First, I am going to explain the basic principle of complex numbers. Our physical interpretation of different elements in complex number mathematics will be followed in this chapter and ensuing chapters as needed. The interpretations and assertions made do not necessarily apply or accepted in contemporary physics. The analysis is derived and defined on the context of a new model.

Complex number system can be defined as a Cartesian system where the x-axis represents the real value and the y-axis denotes the imaginary part.

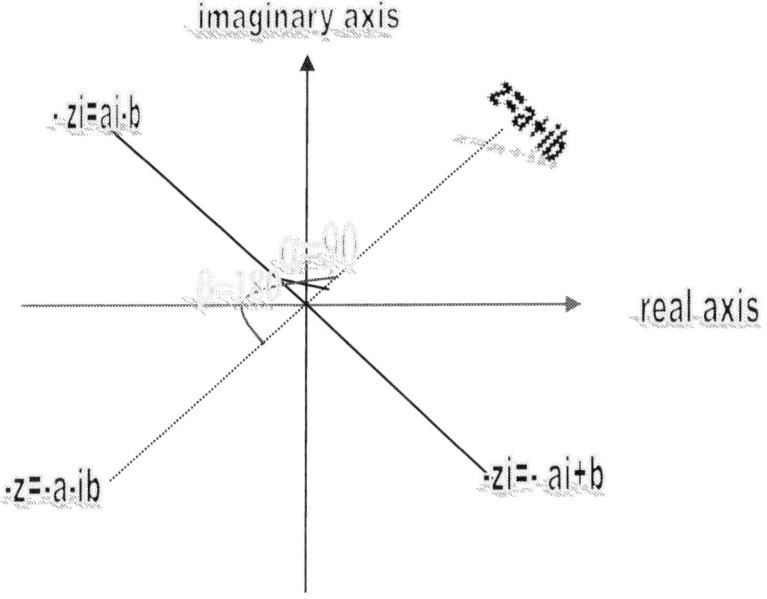

Modified Complex Plane

The second and third quarter of the above diagram is not defined in the current complex number system.

Underneath, the polar system version is shown where $r = |z|$, called the absolute value or modulus, and $a = \arg(z)$, called the complex argument of z.

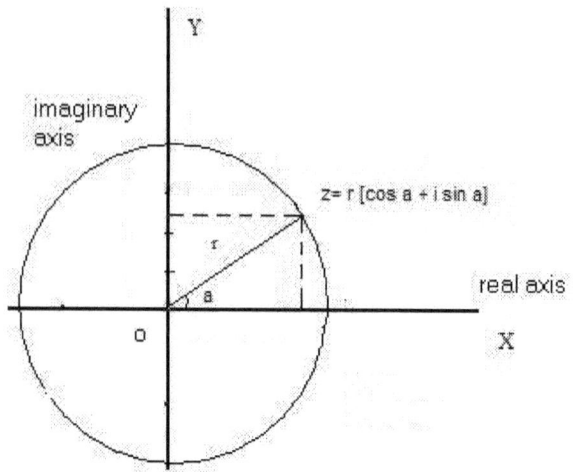

Periodic nature of Complex numbers

The equations for the diagram above can be written as:

Z =x +iy =r (cos,a +i sin,a)

x= r cos a, is called the real part

iy = ir sin a is called the imaginary part.

Does it sound like gibberish? It simply says that complex numbers are a combination of purely real and purely imaginary numbers.

Imaginary Numbers

What are imaginary numbers? In the sixteenth century, Italian mathematicians frequently encountered the square root of negative numbers ($\sqrt{}$-n) in their calculations. Known mathematics could not offer a solution for this problem because nothing could be squared to -1. Eventually, an imagined number was chosen. The square of this imaginary number would be -1. This number was called imaginary number and is represented by the symbol *i*. Later on, it was noticed that a combination of real number and imaginary number is essential to explain the fundamentals of mathematics. This combination is called the complex number system.

Complex number = [x (real number) + iy (imaginary portion)]

In 1806, Jean-Robert Argand, trying to give geometrical visualization to complex numbers suggested the diagram below:

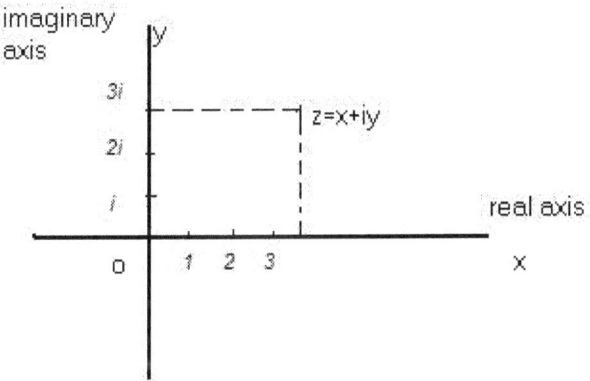

Argand Diagram

In 1799, Gauss proved the fundamental theorem of algebra using complex numbers. Nowadays, the use of complex numbers pervades a major portion of mathematics and its applications in science. Even pure real numbers are rightfully reintroduced as:

N = x + 0i

Where N denotes a number, x is the real value and i conveys the imaginary aspect of it.

This configuration demonstrates the imaginary dimension embedded in any real number.

Please, do not be disappointed if the definitions presented in this chapter do not exactly match the conventional definitions for the complex number system. The descriptions and assumptions made in this chapter are defined within the context of the proposed model in this book. As long as the assertions are mathematically sound, we should be able to rely on them and refer to them in future quotations.

One would assume if imaginary numbers are so fundamental in mathematics, they would have to represent a physical reality. As Roger Penrose points out,

"These strange numbers also play an extraordinary and very basic role in the operation of the physical universe at its tiniest scales." [56]

In this context, I take the real numbers to represent observables and interpret imaginary numbers as non-observables and qualitative values of physical elements. As mentioned, we measure the elements of space-time by real numbers. The notion of complex numbers implies that any of these elements should have an imaginary dimension in their nature. In other words:

Assertion C1: Any computable in the universe also contains a qualitative non-measurable aspect embedded in it.

Therefore, I conclude that at a profound level, just dealing with objective reality is not enough. To get the whole picture we have to open our scope to include non-observable aspect of physical elements as well.

In 1707, Abraham De Moivre found a similarity between complex numbers and trigonometry. These numbers follow the same rules applied to trigonometric calculations. For example, when we square a complex number we double its phase (angle).

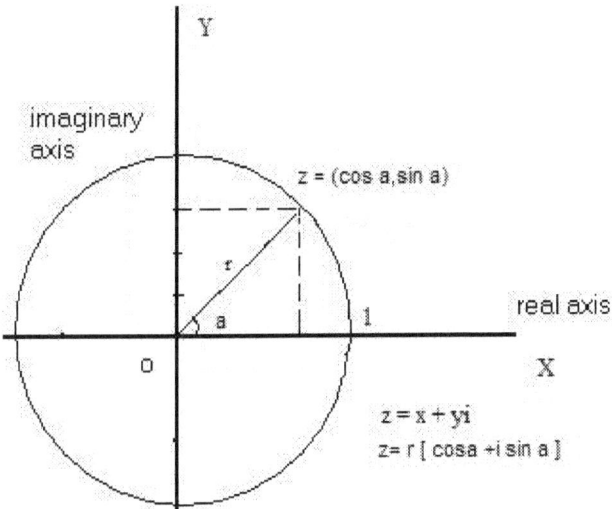

Resemblance of Argand Diagram and Trigonometry

$Z^2 = r[\cos \alpha + i\sin \alpha] \times r[\cos \alpha + i\sin \alpha]$

$Z^2 = r^2[\cos{}^2\alpha - \sin{}^2\alpha + 2 i\cos \alpha\sin \alpha]$

$Z^2 = r^2[(\cos 2\alpha + 1)/2 + i \sin 2\alpha]$

Here, Z^2 is a complex number. $[r^2[(\cos 2\alpha + 1)/2]$ is its real part and $(\sin 2\alpha)$ is its imaginary part.

In this text, we take point zero in the Argand diagram to represent singularity and the imaginary portion of the diagram as the effect of the proposed singularity on the observables.

Imaginary numbers are sometimes called magic numbers. One of the strange characteristics of these numbers is the fact that in De Moivre diagram, any real number coupled with (multiplied by) them will be reduced to zero.

25

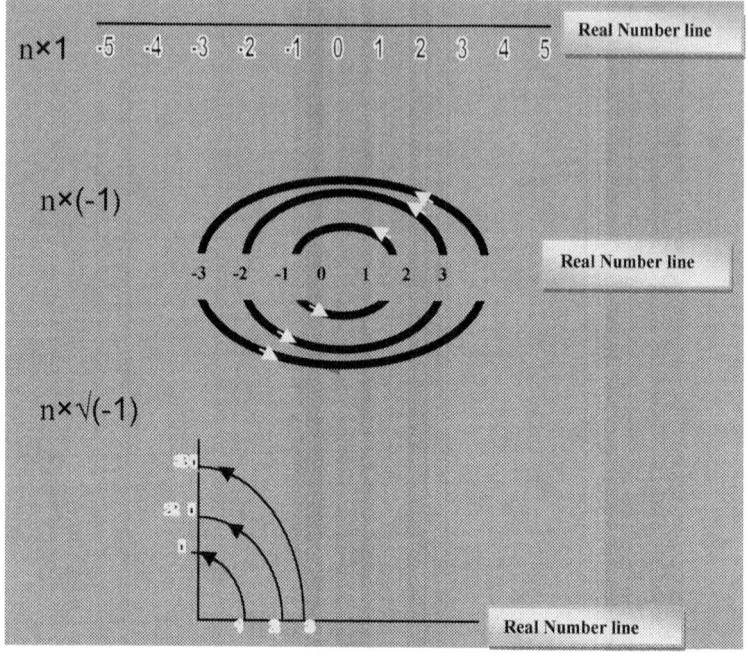

As shown in the diagram above, when we multiply any real quantity by *i*, its real value (its *real number coordinate* value) is reduced to zero. Algebraically this can be written as:

$(X+0i)\ i = Xi + 0(ii) = Xi = 0$

In trigonometry we can show this with;

X=r Cos a, if we take a=90, then Cos a=0, therefore x=0.

Assertion C2: Although the real number field may create the illusion of continuity, the more accurate complex number version shows us that the continuity of real number breaks down periodically.

Therefore, I conclude that physical elements (like space, time and matter) are discrete and not continuous. We can also show this fact by an evaluation of the function of (x) in any equation. We take y = x IxI as an example.

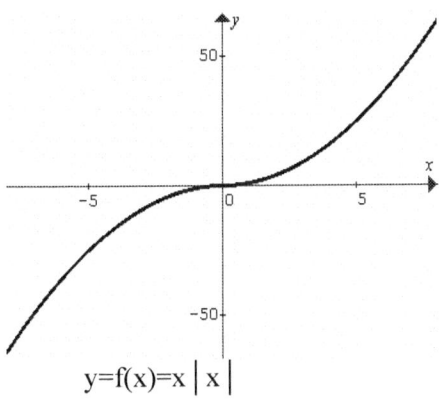

$$y = f(x) = x \left| x \right|$$

Function

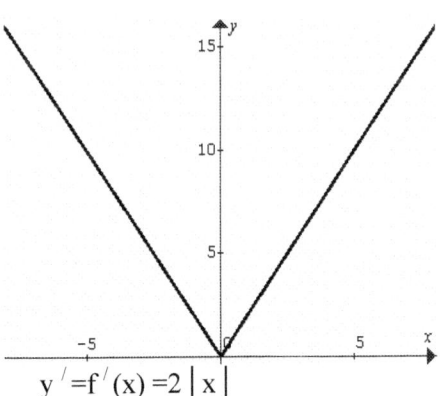

$$y' = f'(x) = 2 \left| x \right|$$

First Derivative

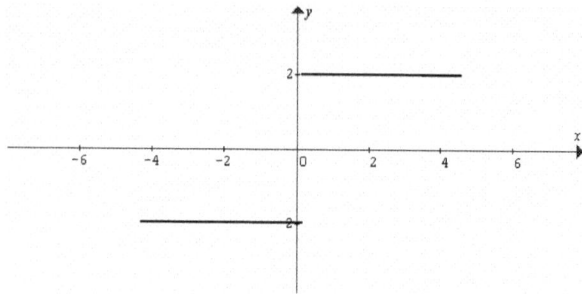

$$y'' = f^{//}(x) = 2 + 4\,\square(x)$$

The second derivative exhibits discontinuity around point zero. A lack of smoothness and continuity in derivatives of real number functions.[56]

For any other function of finite real numbers, we can come to a derivative which shows a lack of smoothness and continuity.

We can take $y = 1/x$ as another example.

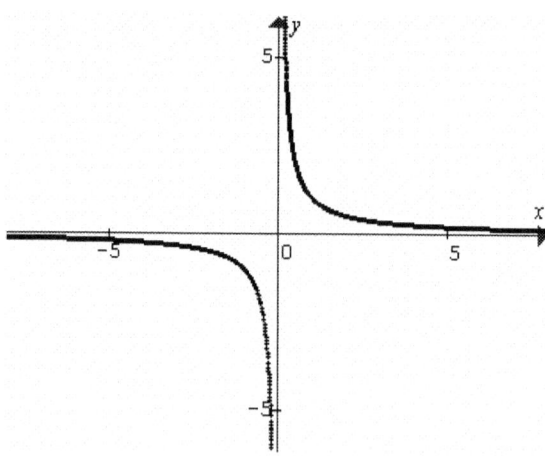

Plot of 1/x

Although the plot is infinitely differentiable, it lacks continuity. The continuity breaks down as it approaches point zero. So real number quantities inherently are not smooth or continuous (holomorphic). If continuity is desired we have to incorporate the imaginary number and adopt the concept of complex numbers into the equation and rewrite the $y = 1/x$ equation as $y = 1/z$. where z is a complex number shown as: $z = x + ib$, where b is any number. The above plots also reveals that discontinuity of real numbers occurs as the curve approaches the y-axis. Therefore:

Assertion C3: The continuity of real numbers always breaks around point zero.

Here, I conclude that as physical elements meet singularity their continuity break down and therefore they are discrete.

On the other hand, we can choose any other point in the domain and shift the point zero to that point and use Cauchy formula in the origin shifted form.

$n!/2pi \int f(z)/(z-p)n+1dz = f(p)$,

And the nth-derivative expression would be:

$n!/2pi \int f(z)/(z-p)n+1dz = f(n)(p)$,

Roger Penrose writes: "Thus complex smoothness implies analyticity (holomorphicity) at every point of the domain."[56]

Assertion C4: The mathematics of complex numbers also indicates that any point in the domain can be considered point zero (cross section of coordinates). This is important for us when we define the proposed singularity and its relation with space-time in the next chapter. Taking every point of the domain as zero is called blowing up the origin.

The complex number equation $Z = R [\cos a + i \sin a]$ indicates that these numbers have a periodic nature. So they loose their real number value and hit zero twice in each period. We take the periodic nature and intermittent appearance and disappearance of real value of the measurable the basis for our fifth assertion.

Assertion C5: Any measurable in our universe has a discrete nature. This will include matter, time and space. For example in the diagram at page 22, if x-coordinate denotes the mass of particles, somewhere in its endeavor the tangible mass gradually looses its value and disappears. On the other hand, if x indicates dimension and distance, because of the function of the complex system, it has to disappear and reappear during each period. This is the basis for our assumption that space and time are not continuum. They have to be discrete.

Another interesting characteristic of imaginary numbers is the fact that although they are influencing the real numbers in equations, they normally do not mix up with them. In a complex number, we normally have to deal with each portion separately. For example for addition we write the equation as:

$$(6 + 3\,i) + (5 + 2\,i) = 11 + 5\,i$$

Assertion C6: The real numbers and imaginary numbers represent two separate domains.

On the other hand, pure real numbers are represented by just one line of the field (X Axis), whereas anything else in the field has a complex nature. This concept can be shown by using the Riemann Sphere. Riemann sphere is an extended complex plane that includes infinity. If we let the sphere represent a physical object, Just one circle plots the pure quantitative value of that object (blue circle for instance). However, there are infinite circles on the sphere that demonstrate the qualitative measures of the object as well.

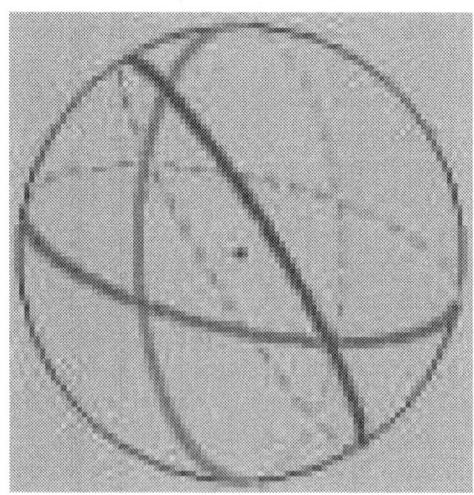

Riemann Sphere

So we may come to conclusion that,

Assertion C7: The real value (computable value) of an object is numbered whereas, non-computable quality of things are infinite.

As mentioned above, in this model, we take the imaginary number (*i*) as a factor, which represents the singularity effect on different phenomenon. Point zero on the other hand, represents singularity.

Similarly, in Cartesian coordinate system if we take x, y and z to represent different values in space-time like, distance, permeability, temperature, weight and so on, we notice that zero is located at the center of all of those values (At the center of the line representing each value) But its not any of them (they are measured zero at point zero).

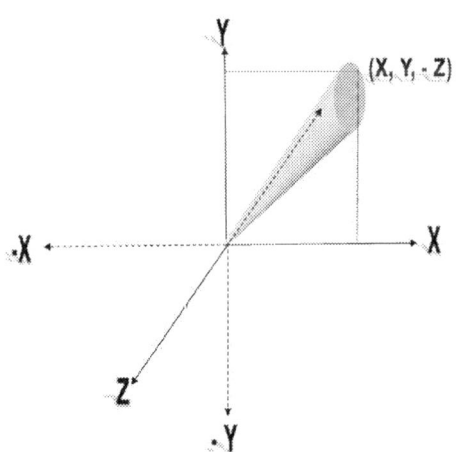

X-axis can represent time while Y-axis represents mass and z - axis may represent volume. Zero is in the center of all of these values.

31

Assertion C8: While point zero is at the center of any value in space-time, it does not represent any of them (values turn to zero at point zero).

Summary

The concept of complex numbers opens our eyes to the combination of the qualitative and quantitative characteristics of the elements present in the universe. Although the mathematics of these numbers is highly developed, the physical interpretation of the complex system is not fully understood. In this model, we hypothesize some physical interpretations for the elements of the system and will examine if these interpretations will offer solutions to existing paradoxes.

My assumptions are as follows. Zero is at the center of the universe. However, it is a separate entity and the value of space-time elements in it is zero. Nevertheless, it affects the space-time and the whole existence revolves around it (unit circle concept). Furthermore, zero exists at any point at any domain and field. That defies the concept of continuity and makes any physical element in our universe discrete.

Singularity
A Separate Entity

"We must try to understand the beginning of the universe on the basis of science. It may be a task beyond our power, but we should at least make the attempt" Steven Hawking

So far, the efforts of the scientific community have been confined to reveal the secrets of our material world. We have progressed so far. At the dawn of the 21st century, every day a new and amazing discovery is brought forward. Each discovery though, presents to us even more puzzles and questions. In physics, the dilemma between Einstein's General Relativity and quantum mechanics (the mechanics of subatomic particles) has attracted much of the effort and research of theoretical physicists throughout the world. Through their endless efforts, the picture of our world is becoming more and more understandable and interesting. In this spirit, I would like to submit the following concepts for enthusiasts to review and examine. If it proves to be based on sound principles it can shed light onto many unsolved mysteries, which are facing us today. If it does not, we could nevertheless consider it as food for thought.

In this book, I am introducing a model in which an assumed version of singularity acts like a medium. By contemporary physics definition, singularity is the zero size ultra dense point, which according to the Big Bang Theory, is the origin of our universe. In this model, I will assume that a burst of energy from a singularity created the Big Bang. However, the singularity itself remains active and plays a major role in the course of action and evolution of the cosmos. The singularity acts like a medium, holds, and connects the space-time universe components.

The idea of introducing a fabric for wholeness of the world is nothing new. The scientific findings provide numerous evidences, which support common sense, and every day experience that the world is an interconnected system. Many attempts to prove the wholeness has been tried which was not completely successful and could not get broad scientific approval. Even The great twentieth century physicist David Bohem tried to present his "Implicate Order" as a model, but unfortunately, he could not finish his Unbroken Wholeness Theory before he died. He believed that "at some deeper level of reality such particles (subatomic particles) are not individual entities, but are actually extension of the same fundamental something."[58]

This is another attempt to present a model for the wholeness of the world. This model is based on redefining the singularity and using it as the background for space-time universe.

Revisiting Singularity

Singularity was brought to our attention after Albert Einstein presented his Field Equations on November 18th, 1915. But Einstein himself was trying to deny it for the rest of his life. In 1939, he tried to show that "Schwarchild singularities do not exist in physical reality."[4]

Then singularity was taken as the nucleus for the initial explosion in the Big Bang Theory. The Theory of Big Bang suggests that the universe started with a huge and rapid expansion of a singular zero size condensed point. However, encounters of cosmology with singularity do not stop there. By definition, singularity also exists at the center of each black hole. Black holes are spots in space where gravity is enormous. It will swallow any object, which comes close to it. Even space and time will be curved and sucked in by this strange phenomenon. Supposedly, there is a black hole at the center of each galaxy.

Black hole with a singularity at its center

In addition, while working with space-time theories this singular point frequently is encountered while we do not have any interpretation for it.

In traditional physics, singularity has been labeled as a 'catastrophe', a problem, a non-real something or simply 'non-physical'. However singularity clearly presents itself in every astrophysical theory, as well as quantum mechanics, and all mathematical equations related to these theories.

As mentioned before, several attempts have been made to bypass singularities. Theories such as Super Symmetry, Super Gravity, Super String etc. were primarily developed to remove singularities and infinities from matter and fields. Theoretical physicists have been trying to ignore it for almost a century;

Brian Greene, one of the main advocates of the Super String theory, declares that while developing the theory,

"We knew that there were significant aspects that we would need to work out before we could establish that our second half of the story did not introduce any singularities – that is, pernicious and physically unacceptable consequences."[1]

But others such as John Earman see it otherwise:

"(This book is) written in the faith that, if adequately revealed, the problem of space-time singularities will not remain the orphan of the philosophy of science and that if adopted as a rightful child, it will enrich not only the philosophy of space and time, but other members of the family as well."[4]

To present an alternative view for singularity, I would like to refer to Einstein's remarks as the gateway for this new definition:

"If we imagine the gravitational field... to be removed, there does not remain a space of type (1) {Minkowski space-time}, but absolutely nothing, and also no 'topological space." (Einstein 1961)

John Earman argues:

"Einstein is surely right that, whatever the technical details of the definition of space-time singularities, it should follow that physical laws, in so far as they presuppose space and time, are violated or, perhaps more accurately, do not make sense at singularities. This is a good reason for holding that singularities are not part of space time."[4]

According to the above, we should forget about space-time as an adjective for singularity, rather we have to consider singularity as a separate entity. Assertion C6 in the previous chapter also proclaims that imaginary numbers belong to a separate domain. From here on in, I also intend to refer to it, as a singular noun, rather than plural. So we call it singularity instead of the space-time singularities.

In his recent book, The Road to Reality, Roger Penrose writes:

"Unacceptable singularities in a classical theory do not necessarily tell us that such blemishes will persist in the appropriate quantum version of that theory."[56]

He then bring the electron as an example where in classical theory it has to spiral into the nucleolus and create catastrophic instability for the ordinary classical atom, whereas quantum mechanics provide solutions for a quantum mechanical atom to stay stable which is in line with observations.

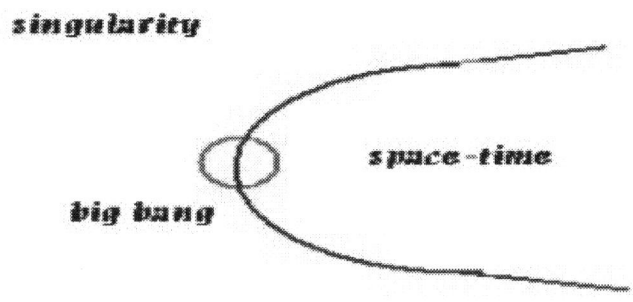

Space-time universe started and evolved from a zero size point called singularity

In this text, I will propose an alternative physics where the singularity is not part of space-time. With this assumption we expose ourselves to a new definition for reality. Let us imagine the singularity as a separate entity out of our space-time universe. This assumption will raise two major questions.

At first glance, one can ask, if our physical knowledge is derived from laws of space-time, how can we define and explain an entity which is out of this physical universe? With this belief, singularity looks abstract. Steven Weinberg, one of the great physicists of this century, questions:

"How can we get the ideas we need to formulate a truly fundamental theory, when this theory is meant to describe a realm where all intuitions derived from life in space-time become inapplicable?"[49]

On the contrary, this model advocates that life and everything else in space-time is influenced and intermingled with that realm. So I assert that our intuitions are applicable and helpful to investigate singularity. Singularity may not be incomprehensible after all.

In addition, we are also facing the quantum behavior that is far from our comprehension with the conventional logic and knowledge. At this time nature and the origin of mind are also obscure. Does that mean that we should abandon the search for understanding these concepts?

Secondly, one may question that, anything out of the boundaries of our universe is so remote that it cannot have anything to do with our lives or our universe. Why should we bother spending time on it at all?

Assertion C4 in the complex number chapter assumes that point zero exists in every spot of the domain. At least in one theory (Loop Quantum Theory) the boundaries of the universe are believed to exist in every minuscule of space. The theory states that space is not a continuum; rather it has a discrete and fabric-like texture.

Assertion C5 also points to the discreteness of space and time and any other computable.

According to above we may be exposed to out of space in every miniscule of space-time. In this paper, I hope, I can show that the singularity is also involved in every basic function in space-time. So Singularity may not be as remote and irrelevant as one may assume.

As mentioned above physicists avoid the concept of a *being* out of the space-time universe, because it is considered incomprehensible and out of reach. I believe if we look closer, we come to realize that some characteristics of singularity are within our grasp.

Cartesian coordinate system

In mathematics, the Cartesian coordinate system is widely used to determine the position of points on a plane. In physics, such points are representing the magnitudes of real world values.

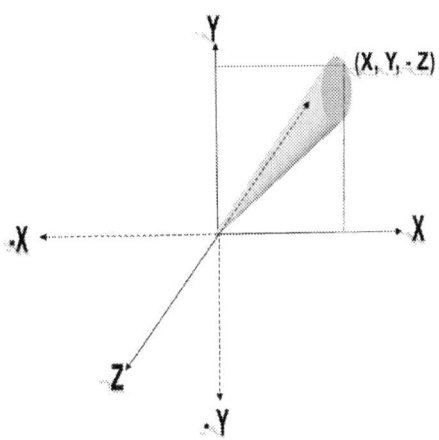

Courtesy of Nahid Sahel Gozin

As mentioned before, in this system zero is positioned in the middle of x coordinate. But at zero point, the magnitude of any quantity that coordinate x may represent in the actual world is zero. Therefore, point zero does not contain or represent x. Zero is also positioned in the center of any other coordinate which may represent other values in space –time. Therefore, we may conclude that, although zero is positioned at the center of any computable in space-time, it does not posses those identities. This is the basis for our assumption that singularity, which is represented by zero in my model, is a separate entity.

Assumption S1: Singularity is a separate entity.

To further define singularity, let us discuss the different properties, which we can or cannot relate to it. Specifically, I am going to mention six elements, whose presence in the singularity can be explored.

39

1)Matter

[Very] Brief History Of The Universe

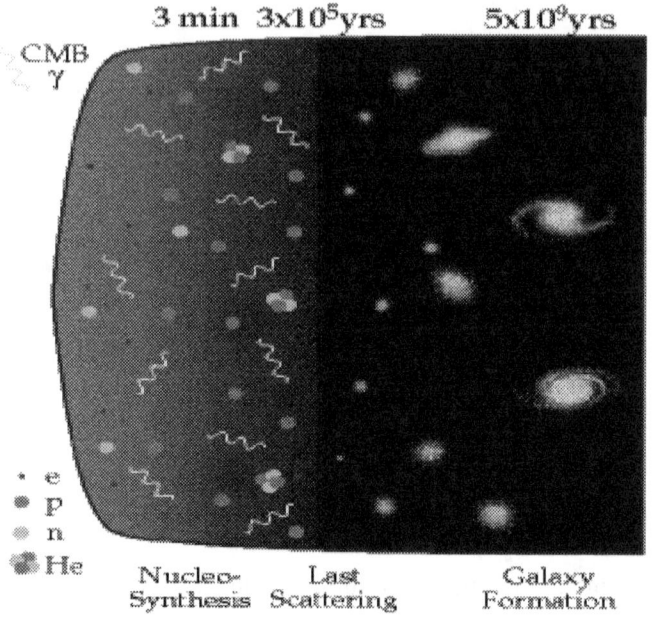

Mass production and the evolution of it after Big Bang

The building block of matter is what we call "thing". By definition the "thing" has zero size and is mass-less. The "thing" acquires mass by traveling through space. The concept is explored further in (Mass & gravity chapter).

The confusion begins when we refer to a singularity as an ultra dense mass. Mass is a property which the "thing" obtains inside the space-time universe.

The most popular scenario of Big bang theory "the Guth's inflationary theory" just talks about rapid expansion of space-time and burst of energy. Is there a period of time before initial rapid expansion of universe? Not much work has been done on before inflationary era and we are not sure if there has been such an era at all. The inflationary theory implies that basic matter particles first appeared after universe expanded, and cooled down a bit. The common particles including, neutrinos, electrons and quarks appeared after the burst. The estimated particles produced at the time is 10^{80}

According to most popular theories, mass appears while the "thing" travels in space. If we define singularity as the origin of our universe – but out of our space-time structure – we have to take it as a no-mass entity.

We can show the fact by using the effect of speed on the mass of an object (Lorentz transformation):

$$m = m_0/\sqrt{(1 - v^2/c^2)}$$

Here m_0 is the rest mass-when the object is stationary-, m is the mass of the object in motion, v is the speed and c is the speed of light. The effect of speed on the notion of time is obtained from:

$$t = t_0/\sqrt{(1 - v^2/c^2)}$$

where t represents time. If we take m as mass of particles inside space-time after the Big Bang (the time that space started and motion could be possible) and m_0 as mass that was present at the beginning of time (t_0), we come to the following conclusion:[40]

$$m / m_0 = t / t_0$$
$$m_0 = m\, t_0 / t$$
$$m_0 = 0 / t = 0$$

Therefore, we can conclude that the mass at time equals zero (time of the Big Bang) had to be zero. Or, we may say that singularity had to be mass-less. We may further conclude that where there is no time (beyond Planck time - smallest amount of time possible), mass does not exist.

In contrast, if we take the singularity as a compressed matter_ with the current definition of matter-, then gravity from such an immense mass would stop the inflation and creation of the universe. Paul Davies argues that in a state of maximum compression of matter, we need "some sort of outward force to overcome the enormous gravity, otherwise gravity would win, and the material would be still more compressed."[7]

He then concludes that under conditions of extreme compression such as what occurred during the Big Bang, there is no force in the universe capable of beating off the crushing power of gravity. Elsewhere, John Earman suggests: "Perhaps. In entailing singular behavior General Theory of Relativity is committed to empirically false predictions"[4]

Inflationary theory predicts that initial enormous expansion of the universe occurred because of a repulsive -negative- gravity. Even so the matter has to be absent at the initial moment.

According to the Big Bang Theory, The simplest form of matter (quarks) first appeared after cosmic inflation. The vacuum energy transformed itself into particles and anti-particles of matter in equal number. There is not clear evidence that at the beginning of time mass has been present. With the ultra-dense mass model, we have to assume that the matter turned to pure energy before the reformation of mass particles. We can as well assume that universe started with a burst of energy and expansion of space. With this assumption, starting point does not have to contain matter.

In such a scenario, we will not have positive gravity force and/or event horizons for singularity. If we remove mass from singularity, it becomes benign.

Furthermore, Assertion C3 in the previous chapter implies that any measurable dissolves as it approaches point zero. We have assumed that zero is representing singularity. Therefore, we may further conclude that the real value of matter has to disappear at singularity.

Assumption S2; Singularity does not contain matter (with its common definition).

2) Space and Dimension

What is space? We know that it is a fundamental structure of the universe. It gives us the sense of locality. We also can sense that it has three dimensions where objects can move along them.

The General Theory of Relativity assumes space and time as real entities that are the background and benchmark of the universe. However, according to General Relativity, these entities are not rigid. Space and time are bendable and play a very active rule in the universe.

Einstein mentions that singularity cannot contain topological space. It means there is no spatial dimension in singularity. In other words, singularity is a mathematical point. Steven Hawking, George Ellis, and Roger Penrose the British astrophysicists and mathematician worked on the Theory of Relativity and its implications regarding the notion of time.

"In 1968 and 1970, they published papers in which they extended Einstein's Theory of General Relativity to include measurements of time and space. According to their calculations, time and space had a finite beginning that corresponded to the origin of matter and energy. The singularity didn't appear *in* space; rather, space began inside of the singularity."[59]

According to the Big Bang Theory, space started at time 0 and has been expanding ever since. Inside the inescapable zone of black holes (event horizon) space gets twisted and disappears. Heisenberg Uncertainty Principle (to be discussed in Quantum Mechanic Chapter) suggests that, in the small scale, location is blurred. Therefore, in ultra small scales, the notion of space disintegrates. We will discuss this in further detail later.

The following arguments support the assertion that space did not exist in singularity. If we deny that mass was the origin of universe we have to consider its energy equivalent as the initiator. The equation for the enormous energy which built our space-time universe can be written as:

$$\psi (E) = E \max e^{i(\kappa r^n - \omega t)} = E \max e^{i(\kappa nr - \omega t)}$$

Where $\psi (E)$ is energy function, i is imaginary number, κ is wave number, nr is number of dimensions, ω is angular velocity and t is time.

$$\psi (E) = E \max e^{i \kappa nr} e^{-i \omega t}$$

In singularity $t = 0$, then $e^{-i \omega t} = e^{0} = 1$
Therefore, $\psi (E) = E \max e^{i \kappa nr}$

If we take energy as maximum in singularity, then $\psi (E) = E \max$ therefore $e^{i \kappa nr} = 1$

If ikrn = 0 we may conclude that n = 0 which means there are no dimension in singularity. Consequently, we can claim that the singularity does not contain space.[40]

Furthermore, Assertion C3 in complex numbers implies that space as a measurable cannot exist in singularity. Also, in the Super Symmetry Model, the size of super-space dimensions which points to the entity beyond familiar space is zero.

Assumption S3: Space is not a property of singularity.

3) Time

The notion of time is a mystery. While space is an objective reality, time is not tangible. A clock actually measures how a sequence of events happens in space. That is how we measure time. Days are produced by rotation of our planet around its axis. Seasons are brought about by different angulations of earth axis while orbiting around the sun.

Besides, everyday experiences direct us to believe that time has an arrow and it is the same for all observers. Both of these notions are debatable.[61]

We know from Einstein's Special Theory of Relativity that space and time are not concrete; rather, they are flexible. About the notion of time and singularity, John Earman writes:

"As Einstein said, physical laws break down at space-time singularities, and for the Big Bang and big crunch they break down so strongly that it is physically meaningless to talk about before and after"[4]

According to the Big Bang Theory, the notion of time does not exist in singularity. Time is a property of space-time universe. In the energy-time version of the Heisenberg Uncertainty Principle (to be discussed later), time gets blurry in the small scale.

And last but not least, the assertion C3 in the previous chapter also indicates that time as a computable element cannot exist in singularity.

Assumption S4: Singularity is not time-bound.

So far, we have discussed elements that cannot exist in singularity and therefore represent the notion of zero. What are the elements that can exists in this domain? Next, I will mention a few.

4) Energy

Energy is the potential for creating change in objects or fields. There are different forms of energy such as kinetic, thermal, chemical, etc.....

If we deny that the ultra dense matter was the origin of our universe then we have to substitute it with another source. The inflationary theory suggests that our universe started with a tremendous burst of energy. This initiated the primary rapid inflation. The followings supports the assumption that singularity contains energy.

The Heisenberg Uncertainty Principle (quantum mechanic chapter) can be extended to the energy of a particle in an ultra short time-span. The Heisenberg equation is written as:

$\Delta E * \Delta T \geq h/2\pi$

Where h is (Planck Constant), ΔE is the uncertainty in energy and ΔT is the uncertainty in time.

$\Delta E \geq h/2\pi \Delta T$. If the product of $\Delta T \Delta E$ is minimum, then we can write $\Delta E = h/2\pi \Delta T$

If ΔT approaches zero then, $\Delta E = h/0 = \infty$

ΔT approaching 0 means the time span has shrunk to ultra small scales and beyond. In this scale, the Energy span approaches infinity. This suggests that the energy itself at the vicinity of the time zero can increase up to infinity

Later on we assert that we are exposed to singularity beyond each Planck time (the smallest time increment possible).

Above, we claimed that singularity contains infinite amount of energy while neighboring our space-time universe, (next to each Planck Distance or smallest increment of space). Shouldn't this huge amount of energy destroy everything just like what is happening in the center of a Black Hole? Isn't Brian Greene right to call it pernicious and a source for physically unacceptable consequences?

Later on, we will see that if a blast of energy lasts for an ultra short time and disappears, it would have no effect in our material world. In the ordinary fabric of space-time, the exposure to singularity is 10^{-43} second at a time (This will be discussed in Chapter 5).

Singularity inside a Black Hole

Inside the Event Horizon of a Black Hole, the fabric of space and time is distorted. So exposure is much more.

"A black hole singularity can be appreciated both as the ultimate garbage dump, able to take care of any waste disposal problem without the need to recycle, and as a source of extractable energy" (Wald 1984a, p. 324-330)

We may assume that the energy inside singularity is vibrant and is the origin for the fields in space-time (for example, electromagnetic field).

We may also use Hawking's formula (introduced by the great British physicist Stephen Hawking) for the black hole's temperature as another indication of presence of infinite energy in the singularity:

$T= hc3/8\pi kGM$

Where h stands for Planck constant, c for speed of light, K for Boltzmann's constant, G for Newton's gravitational constant and M stands for mass.

In the above formula, if mass (M) is zero, the temperature would amount to infinity. Of course, here with lack of matter, temperature relates to internal energy of the system or energy of singularity in this view.

Stephen Hawking says:

"The uncertainty principle of quantum theory means that fields are always fluctuating up and down even in apparently empty space, and have an energy density that is infinite." [6]

Zero Point Possesses Energy

Here is more evidence for the presence of energy in zero fields. Lamb shift and Casimir force are proof for the presence of energy in point zero. If you consider how many zero points exist in the universe, you will appreciate the incredible source of energy that can be derived or is being derived from point zero. Robert Forward[4] at Hughes Research Laboratories in Malibu, California has shown in a paper that at least in principle we can extract the point zero energy to produce electricity.

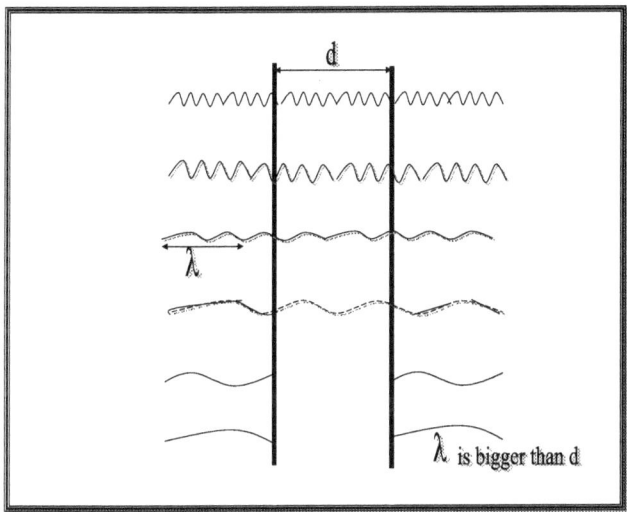

Casimir Effect

Every point of empty space contains energy fluctuation. Two parallel and closely placed metal plates eliminate some of the energy fluctuations. Therefore, the energy of space in between the plates is less than the energy of the outside space. The produced negative pressure then attracts the two plates towards each other.

Courtesy of Nahid Sahel Gozin

Dr. Forward shows in his derivations that, when equations of quantum mechanics are used to determine the average energy (with a bracket on both sides of the E) of the vibrations of the atoms, the answer is:

$E = n(T) + hf/2.$

In a one dimensional harmonic oscillator, the energy levels are quantized. The formula for energy state is:

$E\ n = (n +1/2)\ h/2\ \pi\ *w$
$E\ n =n*\ h/2\ \pi*\ w+ \frac{1}{2}*\ h/2\ \pi\ *w$
$w= 2\ \pi\ f$
$E\ n = n\ h/2\ \pi*\ 2\ \pi\ f + 1/2\ *\ h/2\ \pi*\ 2\ \pi\ f$
$En = nhf + 1/2hf$

47

Where n is the quantum number (photon number) and f is frequency of photon which is emitted by exited atom. In above formula, if temperature equals to zero the particles (electrons) have vibration at energy corresponding to the n = 0 state. Then at T = 0, n = 0 because the atom is not excited (it cannot emit any photon) therefore the value of energy in this case is:[40]

E n =0 + hf/2=hf/2

This is the lowest value of energy. Thus, even at zero temperature (or zero kinetic energy) quantum mechanics predicts that each of the atoms will still have an average residual energy (as we can see if we let *n* go to zero) there remains the hf/2. Physicists have been grappling with this for years because there appears to be an infinite amount of energy available if f is allowed to increase without limit. Ever since Casimir predicted it and various other scientists have verified it, this simple equation is really all that is underlying the theory of the zero point fields and zero point fluctuation. The zero point field hypothesis suggests that there are infinite amount of energy in point zero.

Since Zero point energy (ZPE) was first introduced, many unexplained found explanation. The New Scientist Journal (July 1987) in an article named "Why Atoms Don't Collapse") gives an impressive endorsement of the importance of ZPE:

"There is a dynamic equilibrium in which the zero-point energy stabilizes the electron in a set ground-state orbit. It seems that the very stability of matter itself appears to depend on an underlying sea of electromagnetic zero-point energy."[21]

Both theory and experiment have shown that there is a non-thermal radiation in the vacuum and that it persists even if the temperature could be lowered to absolute zero. Therefore, it was simply called the "zero point" radiation. Further proof is evident, as Dr. Forward points out in his tutorial below,

"When physicists cooled helium to within micro degrees of absolute zero it still remained a liquid! Only ZPE can account for the source of energy which is keeping helium from freezing."[21]

This issue has been re-addressed in a recent paper by the same author:

"This time taking into account what has been learned over the years about the effects of zero-point energy. There it is shown that the electron can be seen as continually radiating away its energy as predicted by classical theory, but simultaneously absorbing a compensating amount of energy from the ever-present sea of zero-point energy in which the atom is immersed, and an assumed equilibrium between these two processes leads to the correct values for the parameters known to define the ground-state orbit. Thus the ground-state orbit is set by a dynamic equilibrium in which collapse of the state is prevented by the presence of the zero-point energy. The significance of this observation is that the very stability of matter itself appears to depend on the presence of the underlying sea of electromagnetic zero-point energy."[22]

The common belief about the source of zero point energy is explained below:

"historically there have been two schools of thought:

Existence by fiat as part of the boundary conditions of the universe, or generation by the (quantum-fluctuation) motion of charged particles that constitute matter. A straightforward calculation of the latter possibility has recently been carried out by the author. It was assumed that zero-point fields drive particle motion, and that the sum of particle motions throughout the universe in turn generates the zero-point fields, in the form of a self-regenerating cosmological feedback cycle not unlike a cat chasing its own tail. This self-consistent approach yielded the known zero-point field distribution, thus indicating a dynamic-generation process for the zero-point fields."[22]

The latter concept attributes the zero point energy to the familiar matter particles that exist inside space-time. This kind of reasoning seems logical and objective and offers comfort. On the other hand, the thought of it coming from the boundaries of the universe exposes us to the unknown and scary realm beyond. Therefore, most physicists would prefer to accept the matter origin for the source of zero point energy. Even if the zero point energy is introduced by quantum fluctuation, we would still need a source for that energy. Maybe we could presume a combination of both mechanisms.

The gravity of different stars and planets has been inserting constant force to curve space and time for billions of years. Applying constant force requires an immense amount of energy. Where does this energy come from?

You may ask how zero point is related to singularity. We explore the relation in the "Where is Singularity" chapter, where we assert that we are exposed to singularity in each minuscule fabric of space (Planck Distance).

The researchers in California Institute for Physics and Astrophysics postulate that zero point energy and its related field are responsible for the acquisition of mass and inertia by the massless particles of the standard model (the model that lists and describes the subatomic particles).

The main objection for accepting the presence of the zero point fields is the fact that its effect has not been observed on electromagnetic radiation throughout space. If such a field exists in space-time, it should affect the cosmic or any other radiation in space-time. This is why some physicists prefer to consider the field a virtual field. If we assume that zero point energy exists out of the boundaries of space-time the problem may find a solution.

Later on, I postulate a model where this out of space ZPE can originate the mass and inertia of particles.

And Last but not least, recent NASA observations reveal the presence of a non-zero cosmological Constant (for explanation please check the flatness problem chapter). The non-zero cosmological constant is responsible for accelerated expansion of the world. According to NASA scientists and other astrophysicists the source of this non-zero amount may be the dark energy, which permeates in from empty space, separates the galaxies, and push them away from each other thus is responsible for acceleration of expansion of the world.

Wilkinson Microwave Anisotropy Probe (WMAP)

I will wrap up this part with a concluding quote expressed by NASA after the non-zero cosmological constant was proved to be true by WMAP findings:

"Many cosmologists advocate reviving the cosmological constant term on theoretical grounds. Modern field theory associates this term with the energy density of the vacuum. For this energy density to be comparable to other forms of matter in the universe, it would require new physics: the addition of a cosmological constant term has profound implications for particle physics and our understanding of the fundamental forces of nature."

Assumption S5: Singularity contains infinite amount of energy.

5) Information

A burst of energy was not the only factor that initiated our universe. There were also physical laws that guided and shaped the universe. These laws and information should have been present at the beginning, or nothing could happen. Can't I then conclude that information is also a property of singularity? And that this property had leaked into our universe at the time of the Big Bang?

The concept of zero point energy proposes that there is energy available at zero point. The amount of this energy carried by sub-atomic particles is calculated as (1/2 h). Where h is the Planck Constant, which is equal to $6.64 * 10^{-34}$ joule per second. Further on in this model we take the Planck Constant (h) as a minimum amount of energy delivered from singularity.

Following the above assumption, we may use the underneath equations to show how the information is delivered to space-time. Planck's constant has dimensions of energy multiplied by time. **h = J.s** (joule-seconds). On the other hand, **h=E/f**. which means that the Planck constant has dimension of energy divided by dimension of frequency.

$$F^2) = M (L/T)^2 = ML^2 T^{-2}$$
$$Df = 1/Dt = T^{-1}$$
$$Dh = DE/Df = ML^2 T^{-2} /T^{-1} = M L^2 T^{-1}$$

Where M is mass dimension, L stands for distance dimension and T is dimension of time. Here we may conclude that the properties of mass, distance and time is included in (h), which means information about these basic elements is contained in minimum amount of energy (h) found in zero point.[40]

Above I have shown that the energy in singularity is the maximum energy. Energy always accompanies a field. The field contains information, so again this is another ground for the assumption that singularity contains information.[40]

Please note that in an empty infinite universe movement of one object does not have any meaning. Motion is relative to other objects. Even the spatial position of each object is relativistic. How can an object be aware of another remote body that is moving, compare to it? How are different objects aware of each other's presence?

Time is also relativistic. There must be universal time information in every minuscule of space so that the laws of physics can utilize it and act accordingly.

Where are the laws of physics located? They are present everywhere. Therefore, these commandments should be the property of a media which is accessible throughout the universe. We have assumed the singularity as the medium.

Black Holes and Information

Black holes are formed by the collapse of big stars. The condensed mass generated from the collapse of the star creates such a huge gravity force, that nothing even light rays can escape from it. Please note that space, time, and matter are not the only elements that are swallowed by a black hole. Information about the entering particles also pours into them. There is actually a debate about where this information ends up after entering black holes.

Stephen Hawking believes that black holes evaporate by releasing energy till they get small enough to pop and disappear. The British mathematician Roger Penrose then questions the destiny of the information that had entered into the black hole's singularity. There are three possibilities. It is either lost or stored or returned back to space-time. Information cannot come back because, as we know, there are not a lot of variations of particles coming back to space-time during the evaporation of a black hole. Just the x-ray photons leave the disappearing black hole. Therefore, it is not likely that information is returned to space-time.

Roger Penrose favors the scenario that information is lost, which is a more reasonable conclusion because the particles that have entered the black hole singularity are destroyed and information has to disappear with them.

Then again, this violates the unity and conservation of information law. The first law of thermodynamics "conservation of mass and energy" can be extended to information as well. Therefore, the only option remaining is that information is stored in a nugget left behind after the black hole's evaporation. But Roger Penrose asks what use this trapped information has if there is no particle to adopt it.

The last scenario could be the answer. If we consider singularity as an informational pool, then it can save and exchange information with space-time.

Assumption S6: Singularity is the domain of infinite information potentials.

The above characteristics describe a kind of entity, which I suppose can exist outside of space-time universe. Singularity may not be the best name for this entity, because it may get confused with condensed mass singularity. On the other hand I cannot resist choosing it because otherwise, it perfectly describes the proposed entity. This singularity is not a pernicious one, because it does not contain ultra dense mass.

Shrinking World

Einstein's Special Relativity tells us that when an object is moving relative to a framework, its size shrinks for some one residing in that framework. The amount of shrinkage depends on speed of the object. If we take this principle to extreme, for a framework which is not moving at all, the moving object downsizes to zero.

If the notion of distance is not present in singularity, logically there is no movement for such an entity. Therefore, we can take it fixed and motionless. For such an entity, the size of moving space-time will be reduced to zero.

I have adapted this concept from *The Universal Theory of Relativity* by Roland Michel Tremblay and *Big Shrink* by Richard Quist. It is interesting how the above authors reflected and expressed the non-locality and shrunk world that they sense in their minds to a new model for the universe.

Departing from the Norm

So far, we have tried to follow physical laws to explain the proposed singularity. However, the result is something, which may not be compatible with the physics that we are acquainted with. On the other hand, departing from scientific norm, when evidence dictates it, is nothing new. That is what Einstein did. Quantum mechanics, which are dealing with building block particles of our physical universe, are also a big leap from classical physics.

I would like to bring Kurt Godel's philosophy of mathematics on his works on undecideability as an indication for the possibility of presence of entities beyond our space-time universe.

"He reasoned that there would always be mathematical statements that are true but can never be proved to be true from existing axioms. He envisaged these true statements as therefore already existing – out there – in a platonic domain, beyond our ken." [7]

Throughout this book, we will explore to see if the above definition of singularity can aid us in finding the solution for the unexplained paradoxes that exist in theoretical physics, and in the nature of consciousness.

Summary

In this view, we define singularity as a separate entity from the space-time universe. It lacks space, time and matter and includes infinite amount of energy and information. We can further speculate that energy fields originate in singularity. We also assumed that space-time universe exists and expands inside the proposed singularity.

In following chapters, I will offer explanations for many unexplained phenomena in physics in the basis of proposed model and explore the mystery of mind and its relation to other elements of reality.

In order to regard this concept as more familiar and tangible, allow us now to explore the mind in the next chapter. We have to accept the mind as an inseparable part of our world. Abner Shimony agrees with Roger Penrose that mentality can be treated scientifically. [5]

Paul Davies states:

"I have come to the point of view that mind is conscious awareness of the world. It is not a meaningless and incidental quirk of nature, but absolutely fundamental facet of reality." [7]

Mind is the tool that we use to interpret and recognize the world around us. Let us use this tool to clarify the above concept further.

Mind

A Separate Entity

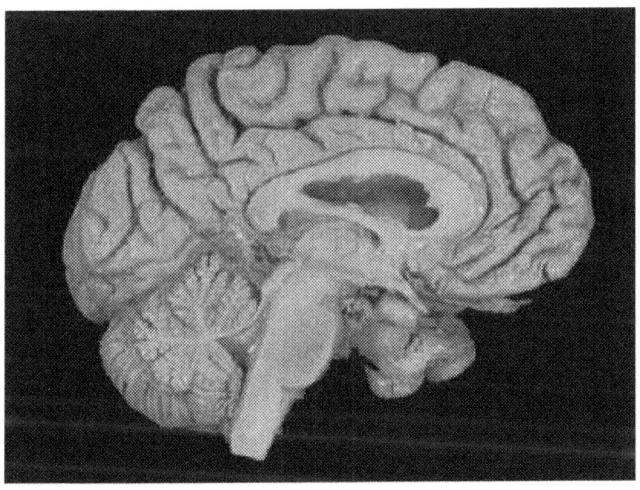

The term mind has such a battered history that it can hardly be used clearly in technical writing, yet it is almost unavoidable. By using the word *mind*, I intend to relate the cognition of an organism about itself and its environment. Consciousness, on the other hand, is a more sophisticated process that higher organisms with more complex nervous systems exhibit in their interaction with the environment. But, for the sake of simplicity, I will use the two words interchangeably.

Plato believed in duality and he described mind as a separate entity from body. Later on, Rene Descartes and his followers further developed this view. Aristotle and later on, Immanuel Kant, Karl Marx and Friedrich Engels in contrast rejected duality. During the past centuries Aristotle's belief that mind is a product of brain function, has been more popular among scientific community. Mainly, because Aristotle's version was seen as more practical and yielding itself to experimental investigations. However, it seems that recent findings and new theories are favoring duality. Santiago Theory of Cognition sees the consciousness as a process and body as the structure.

Francisco Di Biase and Mário Sérgio F. Rocha from International Holistic University, Brasília claim;

> "Consciousness' conception as something essential, primary and irreducible is also found in the consciousness maps, obtained from thousands of psychotherapeutics reports and consistent and converging experiences, observed by several researchers of the medical and psychological areas." (Jung, 1959; Grof, 1985; Moody Jr., 1976; Ring, 1980; Sabom, 1982; Kubler-Ross, 1983; Weiss, 1996)

I adopt duality in this model. Therefore, my first conjecture is:

Assumption M1: Mind is a separate entity from the body.

In following paragraphs I investigate the hypothesis further.

Here again I will use the notion of the complex number system to substantiate the above conjecture.

Complex Number System

As mention before, complex numbers are the basis for today's knowledge. These numbers are a combination of two different entities. Real numbers count for a computable portion of phenomena in space-time and imaginary numbers, which in my view represent the non-computable portion of reality.

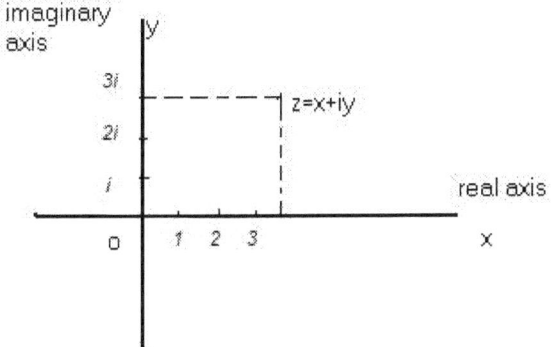

Argand Diagram

In mathematics real numbers alone are just one dimension of the field and in a more fundamental form these numbers are written as,

$$N = X + iy$$

The new form points to the imaginary portion of the field as well.

Assertion C1 indicates that every measurable element has an imaginary dimension to it and as such it implies duality. Since the complex number is fundamental and universally applied in mathematical and physical domains, I would like to conclude that Plato's view is correct and duality is fundamental in our world.

Consensus Reality

The consensus reality (agreeable by all) is the basis for our classical science. The foundation of consensus reality is objectivity and quantitative measurements (which is measured by real numbers). But Dr. Evan Walker[8] believes that upon measuring things by real numbers we are grossly reducing their real value.

By just mentioning a rose, we are not relaying the beauty or the scent of the flower; neither are we expressing its sentimental value and/or the task inherited in it to carry on the legend of rose bushes. In reality, the actual meaning of a rose is much, much more. One can write a whole book about it and the legend even does not end there. Also, the true meaning of a rose differs from person to person. Each of us has a different description or view about a rose bush. The hidden essence in an object is the basis for non-consensus reality.

The qualitative dimension of things is very vast and do not yield themselves to quantitative measurements. This concept can be shown by using the Riemann Sphere. As mentioned before, Riemann sphere is an extended complex plane which includes infinity.

As Assertion C8 states, The Riemann Sphere can represent a physical object where one circle plots the quantitative measure of that object while there are infinite circles on the sphere representing the qualitative value of the object.

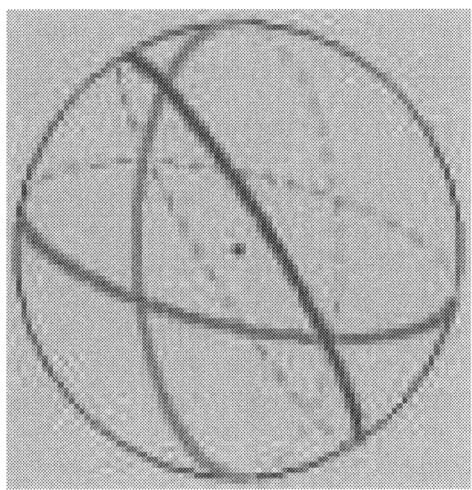

Riemann Sphere

Please note that qualitative properties of objects are in informational domain. They are neither objective nor measurable.

Transpersonal Psychology

Transpersonal psychology is an extension of the humanistic psychology dealing with the expansion of consciousness beyond the usual ego boundaries to include other living things. It does not embrace the limitations of time and space. William James, Sigmund Freud, Carl Jung, Abraham Maslow, and Roberto Assagioli are among the psychologists who have set the stage for transpersonal studies. However one of the most prominent researchers, Stanislav Grof, has carried out decades of clinical studies pertaining to transpersonal psychology. He is a clinical psychiatrist and has held different academic senior positions in John Hopkins University, Maryland Psychiatric Research Center, California Institute of Integral Studies, etc. During his studies of more than 3000 volunteers, he has induced altered states of consciousness by administering medicines such as LSD and by the use of methods like hypnosis, relaxation, meditation, holotropic breathing etc. Many of the volunteers have been psychologists or psychology students themselves.

He describes his findings as being in line with the existence of a "Universal mind and Supercosmic and Metacosmic void."[61]

In his book, Psychology of the Future, Grof classifies the transpersonal experiences of his subjects as follows:

A) Violation of spatial boundaries; in these experiences subjects reported the feeling of oneness with other people, animals, plant life, and with all of creation. There was also a sense of identification with the entire universe. Interestingly, these properties of mind in his studies mimic the non-local characteristic of the proposed singularity (**Assertion S2** in previous chapter).

B) Violation of time boundaries; While most of the subjects had no or limited knowledge about the events in the past, they reported accurate envisions of ancestral, Racial or even planetary evolution and cosmological events (in line with **Assertion S3**).

C) Violation of information Boundaries; the subjects reported detailed conceptualization of organ and body tissues, cells, DNA even the atoms and subatomic particles. Much of this information reported was not prior knowledge to the subjects in their normal state of consciousness. This portion of Grof's findings, concerning the mind, are also consistent with this model, specifically with Assertion S5 in the previous chapter where infinite information is allotted to the proposed singularity. Furthermore, Grof mentions experiences beyond space-time, consensus reality and the ones that he calls "psychoid nature". He concludes "The ultimate of all experiences appears to be identification with Supercosmic and Metacosmic void, the mysterious and emptiness and nothingness that is conscious of itself and is the ultimate cradle of all existence."[61] How better could the proposed singularity be defined! As shown above, Grof's clinical experiences relate the consciousness to a void or as he calls it "nothingness."

Grof also reports that under an altered state of consciousness, his subjects expressed conscious identification with animals, plants and even inorganic materials. Similarly, many pre-industrial cultures believe that "not only the animals, but also the plants, rivers, the mountain, the sun, the moon and the stars appear to be conscious beings."[61] This is also in line with the **Assertion C1** in the complex number chapter which implies that any computable has an accompanying imaginary dimension.

Abraham Maslov, The famous psychologist and educator called mystical episodes reported by his patients "peak experiences". Rather than dismiss such episodes as abnormal, he suggests considering them supra-normal. His experiences show that these events normally lead to "self actualization" and improvement of person involved.

Psychiatrist Walter Planke, describes peak experiences as the state of mind that engenders feelings of...

- unity in the universe (in line with my proposed singularity as a uniting medium)
- strong positive emotion (obtaining energy and insight)
- and disappearance of time and space(in line with assertions S2 & S3 in singularity chapter.)

Grof describes his version of peak experiences, which he calls "experience of the Supracosmic Void", as a state where "it has no specific content, but contains everything in potential form."[61] This is an assumed character which we attributed to singularity (absence of matter and presence of infinite information (Assertions S1 & S5). Transpersonal psychology defies the notion that mind is the product of brain function, rather it considers mind as a universal and extended entity.

Consciousness, a Separate Essence

We have obtained a very detailed map of the brain; we have explored and found neurobiology and biochemistry of different functions of the brain; and we have explored how the brain works; but none of this can prove that the mind is a product of our nervous system.

One might ask, if mind is a separate essence from body, how come animals with more complex nervous systems manifest more advanced consciousness. Shouldn't all exhibit the same complex mind? As an analogy let's look at the motion example, since this example can be applied to the mind as well.

With better, more complex systems, organisms can move better or faster. However the organism does not produce motion, but it adopts it. We may assume that the more complex brain doesn't produce consciousness either, it adopts the awareness. The mind performance of a more complex brain will be superior to a less developed brain just as the moving performance of a species with a more developed neuro-musculature will be superior to that of a less developed species.

Thus we may conclude that our brain is an organ which is actually adopting the consciousness not producing it.

Therefore in this text I adopt the Pluto's duality and make the conjecture that:

Assertion M1: Mind is a separate entity.

Mind as a Tool to Explore Singularity

Sir John Eccles, the Nobel Prize winner for his work on synaptic mechanism, believes a world exists out there that is separate from the material world. It contains all subjective and mental experiences.

By studying mind, we can make the notion of singularity more comprehensible. Try to imagine a *being* with no mass, no spatial dimension, a being that is not time bounded, and at the same time is a main source of information and enormous source of energy. The concept is very unfamiliar and confusing. Or is it?

We have an entity inside each of us that possesses the above characteristics. We call it consciousness or mind. Let us explore the assumed characteristics of singularity in the mind domain.

Mind and Space

While we are sitting in our chair reading these lines, perhaps our mind is elsewhere, in the kitchen and probably thinking about the coffee maker. Perhaps our mind is with our loved one, which is at work or at school, at a distance. If you are fascinated with astronomy, your mind can be two billion light years away at this time. Where is our awareness? Can you show me a location for it? Can you give me a size for it? Is there an actual distance that our mind has to travel to reach to a remote area? May I suggest that the conscious realm does not occupy any spatial position? As far as I know, there is no specific location in central nervous system, which is designated as the center of consciousness. Later on we can see how the Holonomic Brain Theory suggests that information in the brain is stored as frequency rather than a localized bio-physiological process.

Assertion M2: Mind realm is not bounded to locality.

Ein Ajabtar ke mano tou be yeki konj inja
Ham, dar in dam be Araghimo Khorasan mano tou
It is most surprising that we are at this corner
At the same time we are in Iraq (west) and Khorasan (east)
Jalaluddin Rumi 13th Century Persian poet

One can argue that when we are thinking about remote places, our mind is not actually in those places, but uses bits and pieces of memories and information to create virtual locations. Nevertheless, the fact that our mind can be present in different locations in its own domain set the example for a non-local entity. Chris Clarke says:

"Mind is inherently non-local. On the other hand, the world is governed by a quantum physics that is inherently non-local. This is no accident, but a precise correspondence. Mind and quantum operator algebras are the enjoyed and contemplated aspects of the same thing."[9]

Mind and Time

How about time? In our fantasies, we can travel to yesterday, predict events for next week, go back to ancient times, and even travel to the time of the Big Bang. Our mind does not have to wait for the passage of time to finish its journey to remote distances or to time travel. It seems our mind exist in any time or no time. I mean mind is not time bounded.

May I suggest that the rules of time, as we know it in our space-time universe, are not applicable to our mind? For example, unlike real time, which has only one direction, along the history of observer, in our mind realm we can time travel back and forth. Later on, we will see how the notion of time in quantum mechanics mimics the time in our mind because particles can travel in both directions (Feynman diagrams). Except that in quantum mechanics, time is still described by consecution of events that are happening in space. However, in the mind domain, time travel is not defined by the sequence of local events.

Assertion M3: Mind is not time bounded.

Mind and Matter

Matter does not exist in our awareness but its image does. Let us make an imaginary world using the above tools. Just like what we do when we daydream. Suppose you heard on TV that this week's Lottery grand prize is twenty million dollars. You are sitting in your armchair, having your coffee and thinking what you would do if you win the lottery. Of course, at first you would think of paying off your mortgage or car loan and other debts. Then you would start thinking about more exotic things that you could do with that money, like buying a Mercedes Coup, or a mansion with a 50 feet long pool, a tennis court, at least 6 bed rooms. It also would have a very large master bedroom with an on-suite bathroom containing whirlpool, sauna, steam room and any other facility that you may think of. Suppose you are not married yet, you would dream about marrying your ideal spouse and you picture yourself having a very happy life with children in the mansion. Yet, all of the sudden the alarm clock rings. You have to stop dreaming.

Where is that world that you created? Can you pick up a yardstick and actually measure the size of that pool or the land or master bedroom. Show me those children that you were playing with in the backyard. When was, is or will be the date of your wedding? Where is the location of your dream? Doesn't your dream mimic the singularity in the sense that it can contain a big world whereas it does not have any dimension or it is not bound to time? Which one is primary, the dream or the actual mansion?

Isn't it true that dream is the origin of our creativity? Maybe you will stop and buy a lottery ticket in your way to work?

While images of matter are all over our minds, as I mentioned before, no tangible matter could be found in the whole process.

Assertion M4: Mind does not contain the actual material world. But it can embrace an image of it.

Memory

Memory contains information of the material world, while it is not time bounded or does not have any spatial dimension. We are not certain about actual physiological process involved. The Holonomic Brain theory suggests that our memory is restored as spectral frequency not as spatial data.[19] It also suggests that Consciousness has a spectral nature.

No specific location in the central nervous system can be defined as a house for our mind or memory. Memory also exhibits the interesting self-organizing system feature.

Please note that here we are not talking about time elapsed to recall a memory which is a function of neuronal activity and the speed of action potential traveling along nerve fibers. We are talking about the mind realm not the anatomy and physiology of the nervous system.

Mind and Energy

Now that you are here entrenched deeply in this reading, you might crave a cup of coffee, so you put down the book and make a coffee. What happened? What moved you?

For those who practice it, meditation is an energy booster. Our mind is our source of energy and motivation.

Assertion M5: Mind is a source of motivation and energy.

Mind and Information

Mind is the information domain of our body. Information is received, restored, processed and analyzed regularly. According to Carl Jung our awareness also contains more subtle information in a place he calls collective unconsciousness. According to him, our mind has information restored from generations before and through the history of mankind. He also includes information from our ancestor mammals all the way to the first protozoa's and beyond as part of information which is stored in our mind realm.

Assertion M6: Mind is an information domain.

I could go on to explore other characteristics of consciousness; however, I would prefer to stop here and leave you with the fact that our mind does not follow the laws of our space-time universe; rather it mimics the proposed singularity.

While concentrating on the behavior of the mind, we feel a sense of familiarity and acquaintance with the singularity.

Interestingly, mind activity mimics some of the very strange quantum mechanical behavior. Below, we are going to explore some of these similarities.

Superposition of States

By definition, super-position of states refers to the situation whereby an entity exhibits two or more, often contradictory, states at the same time. In our consciousness we are frequently exposed to superposition of states. For example, simultaneous presence of contradictory emotions (love and hate) is our common experience with mind activities. In our mind we can accommodate simultaneously the different possible outcome of shooting towards the Schrodinger's cat (explained in Quantum Mechanics Chapter). In reality, either the bullet hits the cat or it misses and cat escapes. However, in our mind we can superimpose both outcomes. That is how we think, evaluate and plan accordingly. The super position of states in conscious realm mimics the superposition of particle's properties in quantum mechanics.

Decoherence – the blossom of just one state in macrocosm out of infinite quantum superposition states in micro scale.

Critics of the above view, reason that Super position of states in mind is occurring in an informational domain, whereas in quantum mechanics the superposition has materialistic origin. The act of decoherence dissolves the superposition of states at quantum level and changes the super-position to only one objective state observed in the macro-world. This is how it works. The objects are not isolated systems. They are in an environment and in constant interaction with other particles and photons. For example, cosmic rays can interact with the particles in an object and reduce their states to one of the possible state. This is how we do not see an object in a chaotic and superposition condition.

On the contrary, Super-position in mind domain is reduced by the act of decision-making. We cannot therefore equate fully the quantum mechanical state reduction to what happens in mind realm.

Please note that superposition in quantum mechanics is happening in the grey area between virtual and objective world. One could reason that the two phenomena are essentially the same. At the end of process, we end up with just one status and superposition is resolved. The difference is we are dealing with two different domains, (informational and material).

Entanglement

By definition, entanglement is the correlation between two spatially separate objects. Quantum entanglement is one of the main paradoxes in theoretical physics.

Harald Walach a psychologist from University Hospital Freiburg, Germany, extends the prediction of Weak Quantum Theory that entanglement should ensue between parts of a system, to a group of his patients. In his experiment, he chose a group of patients and randomly prescribed a medicine or placebo tablet for them. He shows the improvement of the patients who received placebo treatment compared to patients who actually received the pharmacological treatment was quite high.[78] He relates the improvement of patients who have not received the actual medicine to entanglement. Walach reports many other clinical experiments of himself and other researchers that suggest generalized entanglement between the patients in experiments. He postulates,

"By that mechanism (entanglement) quite a few other hitherto unexplained phenomena, which are deemed unscientific, could be explained. Among them would be relational phenomena as the basis for healing, ritual and other branches of complementary and alternative medicine, or para-psychological phenomena. Consciousness, possibly being itself a variable complimentary to matter, would enter the field by ways yet to be explored and would take a new role."[44]

Dr. Karl Pribram, professor of neuropsychology at Stanford University states:

"We got the brain data first, and then we see, look, it fits the same mathematics as quantum mechanics.....the problems that have been faced in quantum mechanics for the whole century -- well, since the twenties -- and those paradoxes also apply at the psychophysical level and at the neuronal level, and therefore we have to face the same sets of problems. At the same time, I think what David Bohm is doing, is showing that some of the classical conceptions which were thought not to apply at the quantum level, really do apply at the quantum level ...... if indeed we're right that these quantum-like phenomena, or the rules of quantum mechanics, apply all the way through to our psychological processes, to what's going on in the nervous system -- then we have an explanation perhaps, certainly we have a parallel, to the kind of experiences that people have called spiritual experiences. Because the descriptions you get with spiritual experiences seem to parallel the descriptions of quantum physics."[45]

Dr. Pribram like many other researchers testifies for close similarities between mind realm and quantum mechanics.

The many documents available regarding the resemblance between quantum mechanics and the mind are accessible on the Internet or elsewhere. It is enough to say that mind can be better described at sub-atomic and Planck platform rather than from cellular or molecular levels.

Previously I have speculated that mind, quantum world and astrophysics are interconnected and together they reveal reality in a deeper level.

It is astonishing that the above three apparently different topics (mind, astrophysics and quantum mechanics) are interlocking with each other like different pieces of the same puzzle.

Isn't also ironic that our power of creativity is initiated in our mind, and the creation of the universe is initiated in singularity? Roger Sperry has written voluminously on his personal concept of mind and consciousness. In his view mind is one of a whole class of superior properties that emerge from material systems when they achieve a certain high level of organization.[10]

Quantum Mind

Still in doubt? I would like to refer you to a very interesting book, Quantum Mind, by Arnold Mindell a psychologist and physicist. Using scientific methods, he compares the mind with quantum behavior, the mathematic field and astrophysical theories, showing how closely they intermingle and mimic each other to create a whole system. Dr. Mindell demonstrates that the mathematics of quantum physics reveals the pattern behind psychological and spiritual methods of unfolding dreamtime. Francisco Di Biase and Mário Sérgio F. Rocha from International Holistic University, Brasília, the authors of *Information, Self-Organization, and Consciousness: Towards a Holo-Informational Theory of Consciousness* state:

> "Our universe structured as a quantum holo-informational non-local field full of quantum potential with meaning, is an intelligent (informed) universe functioning like a mind, as Sir James Jeans already had observed."[20]

This holographic organization is what the late physicist David Bohm calls implicate order. His model includes, space and time in its structure as an enfolded dimension. Functioning in this holograph mode, our brain can "mathematically builds the objective reality" interpreting frequencies originally from another dimension, from a fundamental order, a holo-informational field located beyond time and space.[20]

> "Our mind is a subsystem of a universal hologram, accessing and interpreting this holographic universe. We are; interactive resonant and harmonic systems with this unbroken self-organizing wholeness. We are this holo-informational field of consciousness, and not observers external to it. The external observer's perspective made us loose the sense and the feeling of unity or supreme identity; generating the immense difficulties we have in understanding that we are one with the whole and not part of it."[20]

We will further discuss this issue at holographic theory chapter.

Collective Consciousness

Carl Jung's collective unconsciousness also assigns a borderless entity to mind. Jung the father of modern psychology states that in deeper level of consciousness, we are aware of our past generation's endeavors during human history. He believed in general sense there is a common awareness in between living things. His idea later developed to *Transpersonal Psychology* by Abraham Maslow, Roger Welsh, Stanislav Grof and others. The educator Robert Hutchins, Ph.D, points out that transpersonal psychology is the future norm of psychology. He believes our personality is the crust or skin covering our transpersonal essence. He further remarks, transpersonal psychology emerges out of personal psychology, as a result of the individual growth and maturation. He also believes that transpersonal psychology asserts that religious and mystical experiences and the perspective that derive from them are valid approaches to reality and can be studied scientifically.

Mind, Mathematics and Physical World

Roger Penrose in his interesting book, *The Large, the Small and Human Mind*, questions the idea that the mind is a product of the physical world. Rather he suggests that the mind and the physical world influence each other.[5]

The creation of the mathematics that deals with actual physical entities can be explained materialistically. One may say that the Human race dealing with material objects has created the mathematics to facilitate his interactions with the outside world. But if our mind developed a math that goes beyond the objective physical world, and if a part of that math can foresee and guide us through unknown territories like quantum mechanics, can we conclude that the mind has its own entity separate from the body that accommodates it?

Long before introducing Planck distance, point was defined as an entity, which does not have any dimension. By experience, we should have described it as the smallest point that a piece of charcoal can create in a stone. But somebody's mind explored deeper into the darkness where our every day experience could not reach. Now we know that the smallest possible size in space is plank distance 10^{-33}cm. Beyond that, we are exposed to geometrical point with no dimension and no size.

Similarly, the negative number era does not have an objective meaning in macrocosm either. However in the 15th century, mathematicians had to include negative numbers to be able to complete calculations. Without entering that new era they would not have been able to close the field since positive numbers were inadequate in completing the mathematical calculations. Later on, the discovery of the electromagnetic revealed that negative numbers are describing a physical reality.

Long before the development of quantum mechanics, mathematical calculation led Nicolas Chuquet an Italian mathematician to the square root of –1.75 in 1484. Of course he disregarded that portion of his calculation as meaningless with no reality. But this portion appeared and reappeared in future calculations of others. The mathematical field again appeared incomplete and ostensibly open.

The square root of minus numbers could not be described with the mathematics of the time. To resolve this, a new arena was needed to complete and close the mathematical field. Gottfried Leibniz and others decided to add imaginary numbers to real numbers as a solution. For years, these imaginary numbers were generally considered unreal and impossible. Even Jerome Cardan who first had to use these numbers to create a formula considered them as meaningless, fictitious and imaginary. But later on, while studying quantum mechanics these numbers proved their authenticity and value. Is mathematics a separate domain out there?

> "For Leibniz, imaginary numbers where a fine and wonderful refuge of the divine spirit. Almost an amphibian between being and non-being...."[14]

In twentieth century, we noticed that we have to use complex numbers, which are a combination of real numbers and imaginary numbers to present a mathematical model and explain the subatomic domain (Quantum Mechanics). In fact mathematical calculations are ahead of us and are guiding us into a deeper understanding of the physical world. What is the origin of our mathematics? Is this convincing enough to accept mind as a separate entity.

Consciousness, a Non-Computable Domain

Any element in objective world is computable. But in mind realm elements are continuous and non-computable. As Roger Penrose states:

> "Non-computability in some aspects of consciousness and, specifically, in mathematical understanding, strongly suggests that non-computability should be a feature of all consciousness"[5]

Roger Penrose then wonders:

"How one goes from computable discrete system (*physical world*) to a continuous system (*mind*)"[5]

In applying the concept to physical reality, one can claim that every concrete noun (like apple, planet, body, etc.) is quantitative and therefore belongs to space-time. Obviously the concrete objects require space and are time bound. On the other hand any abstract noun (like wisdom, hate or joy etc.) is non-computable and therefore belongs to singularity and mind realm. Abstracts nouns are not quantitative and do not occupy any space. They are also not time bound.

Roger Penrose proclaims:

"If there indeed exists some sort of contact with Platonic absolutes which our awareness enable us to achieve, and which cannot be explained in terms of computational behavior, then that seems to me to be an important issue"[5]

In this view, the main building blocks of our physical world: space, time, and matter, all have unbreakable basic units and beyond these limits they fail to exist. Therefore, I conclude that the objective world is a computable entity.

Likewise, if mathematics contains a non-computable entity, we should open our mind to a non-computable aspect for the world as well. This leads us to a kind of duality in existence.

Summary

About theory of everything Roger Penrose states: "...there could easily be a non-computational nature in the correct theory, if we ever find it."[5]

Above, I have examined how actual space, time and matter, does not exist in the mind realm. The representations of such notions in consciousness are merely images. I have also shown how the mind is a source of energy and information, which mimics the characteristics of the proposed singularity, therefore enabling us to assume that the mind may be an extension of the singularity. I also have covered the studies which reveal close connections between mind and quantum mechanics

In addition, I have shown that mathematics has a larger domain comparing to our physical knowledge. Thus, mathematics cannot be explained as being the product of the interaction between the mind and the physical world. In so many instances mathematics is more advanced than our experience. In fact, physicists are following mathematics to offer theories and to explain the unexplained physical phenomena. Finally, I deduced that mathematics emerges from our deeper consciousness.

One of the objectives of this text is to introduce mind as a separate entity and to emphasize that studying and exploring its connections with macrocosm and microcosm (the world in large and small scale) is an essential part of studying reality.

Singularity and Space-Time Universe

Jupiter and Mars in their orbits – Source: NASA

Isaac Newton believed that space is the background where motions are taking place. Therefore, he took space as an actual body and absolute benchmark for the universe. On the contrary, Gottfried Leibnitz believed that heavenly bodies are moving relative to each other, but there is no background. Thus, he denied space as being an actual entity. Ernst Mach had a similar idea, but he introduced the acceleration as a contributing factor for the motion of stars. He believed the relative motion of heavenly bodies is affected by mass distribution throughout the universe.

Albert Einstein however, introduced the combination of space and time as the cosmos background. He declared space and time as actual entities. In his belief, although space and time individually are relative and malleable, the space-time combination builds the fabric of the universe. In his General Relativity Theory, space-time is vibrant and active in the evolving world. The gravity of stars changes the shape of space and curves it. The curved space bends the trajectory of other stars. Hence space-time is not just a rigid and passive background but a dynamic entity.

Einstein's model has passed the test of time and many experiments and observation confirm its precise predictions. Underneath, I will adopt Einstein's space-time and build my model around it.

Singularity and Space-Time

Here, I will assume that space-time universe is enclosed inside the proposed singularity. If this is true, where are we going to find the singularity? Where are the boundaries of space-time? One can speculate, if space-time universe is expanding, does it crystallize and push the singularity away? In this scenario, singularity has to have an internal dimension, which is against our original assumption. Maybe we can assume that our universe dissolves in singularity. But 'dissolve' is not the right word either. Even superposition does not define the actual meaning. At the same time, I am suggesting that, space and time is -in a way- embedded in singularity, so does matter and everything else, which exists in our universe.

The singularity, as defined raises new questions and concerns that we have to offer reasonable answers for.

The first question which may come to mind is:

If singularity is a mathematical point, how can our enormous space-time universe grow inside it?

For the sake of argument one can look at the image of a three dimensional object in a zero dimension (0).

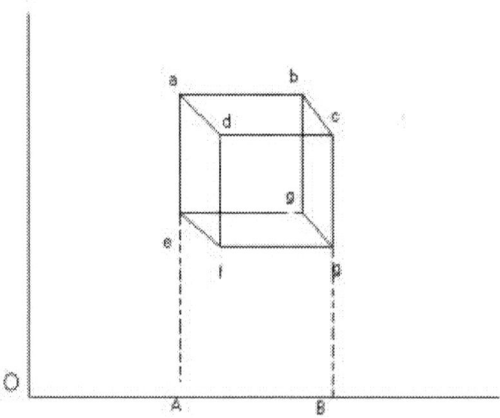

Singularity

As the diagram shows, the image of a three-dimensional object in a two dimensional world (book page, monitor screen) would be reduced. That same image in a one dimension (X-axis) is just a line and in no dimensions (point zero) the image coincides with zero. No matter how big the object is, its image in 0 has no size. In other words, no-dimension (0) can accommodate the image of any size object.

Physical reality of the above can be seen in holography. In holography, a two-dimensional picture can contain all of the information of a three- dimensional object. "Even a tiny fragment of the 2-D holographic plate contains enough information to reconstruct the whole 3-D image of the object."[6]

The algebraic can be written as:

$X * 0 = 0$
Any number times Zero = Zero

As long as we deal with finite numbers, the above equation holds. In space-time we are dealing with finite numbers.

From the above arguments, I want to conclude that if we deliver the universe to singularity, the dimension of it in singularity domain is reduced to zero.

Are the above very simple substantiations? Roger Penrose believes that as we develop deeper and deeper theories, the mathematics become simpler.

Later on, in the discussion about Planck distances, I will suggest a more objective understanding of the whereabouts of singularity. But at this time let our imagination fly and help us to visualize the relation between space-time and singularity. Just suppose you are looking at an object in a mirror. Can you claim any interface between image and mirror? In this scenario, when you look at the image, you will also see the mirror everywhere.

To make this concept more tangible and objective, imagine that you are looking at a piece of sponge with a magnifier. First, we see the roughness of the surface. By increasing the magnification, we can also see empty holes in between. At this point we can claim that the sponge is perforated by many empty holes. Nevertheless this is not an ideal analogy for the correlation of singularity and space-time because empty space requires dimensions whereas the proposed singularity does not possess any dimension. Perhaps the best way to analogize this concept is to imagine the sponge immersed in an infinite entity which cannot be measured or understood with our existing limitations.

Be yeki naghsh bar in khako, bar an nagsh degar
Dar behesht abadio shekarestan mano to
In one mold we are in earth, but in another mold
You and I are in an infinite sweet paradise

Rumi

Real and Natural Numbers

Our world is built based on mathematical fundamentals. Therefore, we frequently look at mathematics to build physical models.

We have two different mathematical choices to adopt in order to build a model for space-time.

If we take real numbers to represent space, we may take the world as a continuum. Real numbers with an infinite amount of decimals (2.567854332234...) stand for continuity in the field.

In contrast, natural numbers (1, 2, 3, 4, 5......) with their discrete nature symbolize discontinuity of the field. If we adopt natural number system as the model then our space-time at fundamental level has to have units and is discrete.

But as it was mentioned earlier the correct mathematics at fundamental level is the mathematics of complex numbers. So perhaps we have to take our model base on the complex system.

Imaginary Numbers

In mathematics, the square root means the sides of a square with area *a*. Imaginary numbers (square root of minus numbers) are frequently being used as an integral part of quantum mechanical mathematics. These numbers are representing the hyperspace (a space beyond 4 familiar dimensions) which is influential in the quantum arena.

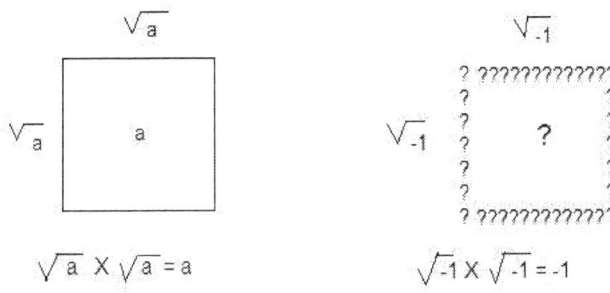

$\sqrt{a}$ X $\sqrt{a}$ = a

Square root of a is the side
of a square with area equal
to a

$\sqrt{-1}$ X $\sqrt{-1}$ = -1

How do we define a geometric figure
with negative area?

By definition, square root of a ($\sqrt{}$-a) is the side of a square with area equal to -a. What kind of a square would have negative sides? It cannot have a meaningful real (tangible) area. It cannot have a meaningful dimension either. So, I want to conclude that contrary to string theory, hyper-space cannot have a real dimension and area. We call the Hyper-space in this model *the singularity* which does not posses any dimension.

Complex Number Mathematics
Now let us go back to complex number system (a two dimensional system which consists of a real and an imaginary coordinates). It has been explained in detail in the first chapter.

Assertion C#3 indicates that discontinuity of real numbers always happen around point zero. One of the strange characteristics of imaginary numbers is the fact that any real number coupled with (multiplied by) them will be reduced to zero.

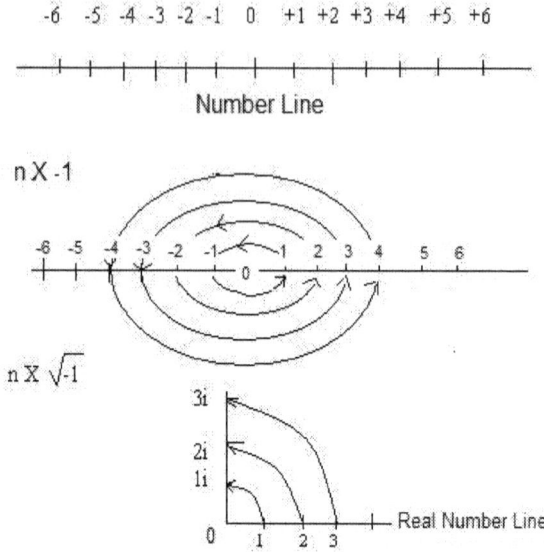

As shown in diagram above when we multiply any real quantity by *i* its real value (Its *real number coordinate* value) is reduced to zero. Its algebraic will be written as:

$(X+0i)\ i = Xi + 0(ii) = Xi$

Therefore, the real value disappears and imaginary value demonstrates itself. In trigonometry we can show the fact as;

$X = r\ Cos\ a$ since we took $a = 90$ and $Cos\ a = 0$, then $X=0$

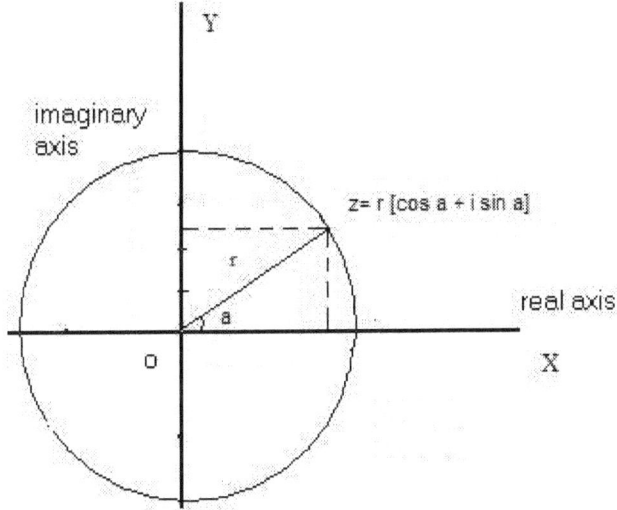

Periodic nature of Complex numbers

The complex number equation $Z = R [\cos a + i \sin a]$ indicates that these numbers also have a periodic nature. So they loose their real number value and hit zero twice in each period which indicates discontinuity in real number field. Therefore, any space-time element (space, time, matter) shown by X coordinate, as it couples (multiplies) by imaginary number, looses and regains its real value periodically. For example, if x indicates dimension and distance, because of the function of the complex system, the space has to disappear and reappear during each period. This is the basis for our assumption that space and time are discrete and not continuous.

Erwin Schrodinger was one of the first physicists who suggested a discrete space. Einstein in his last published paper, having quantum theory in mind, also proposed that a discretely based theory might be the way forward for the future physics.

As it was mentioned before, in this model we take the zero point on the complex number plane to represent singularity. The imaginary number (*i*) represents the singularity effect on different space-time phenomenon.

Assertion C2 indicates that continuity of real number breaks down intermittently. What are we going to get out of these argument?

85

If we take the space-time as being discrete and if in every period the real value hits 0, then we may conclude that space-time is spread over singularity like a web. But then we assumed that the singularity is zero size. How are we going to imagine a web spreading over zero? We may find the answer by noting that the X coordinate denotes the value for dimension in space. Dimension is not defined in singularity. Assertion # S1 denotes that singularity is a separate entity. We are talking about two separate domains. If we cannot expect and assume dimension to singularity, then we have to depart from thinking objectivity as the only tool. Next, we let our mind's eye lead us and imagine a 4-dimensional universe and a 0-dimensional entity intermingling and amalgamating with each other. One can also argue that, singularity is a point, because we are not comparing it with anything of its kind, in its own domain. Keep in mind that we are assessing the singularity with a space-time parameter.

When talking about incomputable parameters, Douglas Hofstader concludes,

> "Undecidable propositions run through mathematics like threads of gristle that criss-cross a steak in such a dense way that they cannot be cut out without the entire steak being destroyed"[7]

What does this actually mean in real world?

Assertion C4 indicates that any point in the plane can be considered point zero. This supports are assumption that point zero is present in proximity of any miniscule of the space-time meshwork. We can further assume that zero is present next to each atom of pace and time.

In mainstream physics considering every point of the domain as zero is called blowing up the origin. This is done with a mind-state that localizes the origin somewhere in space-time. Then spreading zero all over the domain is considered exploding the origin. But in this model origin is a separate entity which is accessible in every point of space-time. So the origin remains intact.

Assertion C6 supports the notion that the singularity and our universe are two separate domains.

On the basis of the above arguments, we may conclude that in the gaps between space-time webs, we are faced with the 0-dimensional singularity.

Super-Space

The need to look beyond the ordinary four-dimensional Minkovsky space-time to explain many physical phenomena has been noticed for some time now. Many phenomena, for example, electromagnetic fields cannot be explained in the context of a four dimensional universe.

To explain these mysteries, main stream physicists chose to theorize another space-like manifold attached to ordinary space-time. This manifold is called super-space. In basic terms the super-space idea presumes that the points in space-time are actually cross sections of bundles which are extended in this proposed super-space.

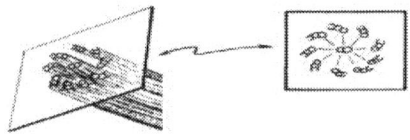

Every point in space-time is a cross section of an extended thing.

Extra dimensions in super space are called internal dimensions. Therefore, every point in space-time has internal dimensions that are out of site. Interestingly these bundles (fibers) are frequently represented by a complex vector one-dimensional space, which include an imaginary portion.

The introduction of super-space opens up a can of worms and exposes theories to the doldrums of extra dimensions. The Super String Theory is proposing up to 7 extra dimensions. Did super-space solve the mysteries? Not quite. On the contrary, it created a lot of chaos and dragged the theoretical physics to places which are far from objectivity. Maybe it is time to forget about extra dimensions. Maybe it is time to think about a non-space like entity adjacent to our universe. For the adjacent entity to space-time, I am proposing the singularity that is a non-spatial essence.

Fabric of the Universe

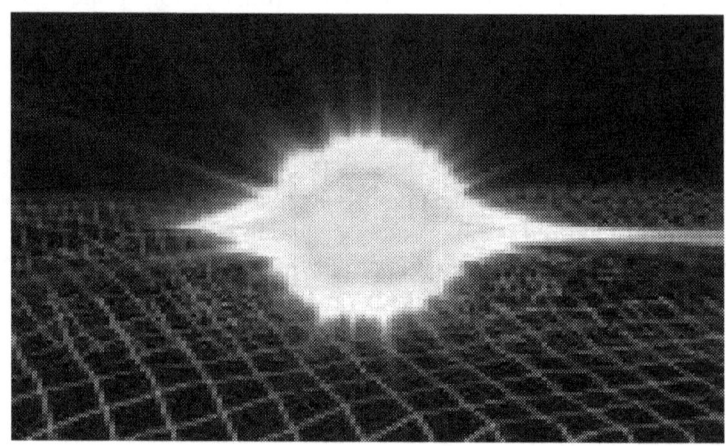

Many schools and physical theories are considering the space as a continuum. But others like Loop Quantum Gravity picture time and space as discrete elements. In presented model, we take space and time as being discrete as well.

Mathematical discontinuity represents holes in the fields. Dividing a real number by zero creates discontinuity in the math domain. This same problem renders itself in the General Theory of Relativity. This gives the astronomers the sense that there should be holes in space-time.

The Planck length is the scale at which classical ideas about gravity and space-time cease to be valid. This is the 'quantum of length', the smallest measurement of length with any meaning. It roughly equals to 1.6×10^{-35} m or about 10^{-20} times the size of a proton.

Inside of the Planck length, the notion of space is not valid anymore. Can we suppose that at the border of each Planck distance our matter-space-time universe ends?

Planck time is the time it would take a photon, traveling at the speed of light, to cross a distance equal to Planck length. This is the 'quantum of time', the smallest measurement of time that has any meaning, and is equal to 10^{-43} seconds. No smaller division of time has any meaning. Within the framework of the laws of physics as we understand them today, we can say only that the universe came into existence when it already had an age of 10^{-43} seconds.

If no distance less than Planck length and no time less than Planck time has any meaning, can we assume that Inside Planck limits is out of our space-time universe, because it does not contain meaningful space or time?

Planck time is the building block of time and, as such, it is the absolute unit of time. With the same token, Planck length is the building block for space and as such is the absolute unit of space. Out of this boundary, the meaning of space and time breaks down. The above arguments substantiate and favor a space-time with discrete fabric over a continuous one.

Exposure to Singularity

We have assumed that universe is contained inside the singularity. Therefore, based on our assumption we can claim that, our exposure to singularity in each centimeter of space is $1.6 * 10^{35}$ absolute units of length. In other words, the exposure is in every minuscule of space, or everywhere.

One also can claim that, in every second we are exposed to naked singularity 10^{43} absolute unit of time. That means all the time.

Formerly, it was assumed that singularity contained merely energy and information and that it is benign. How is it possible to be exposed to infinite energy and not experience catastrophe? The laws of physics tell us that, if blasts of energy were confined to short enough intervals, these blasts would have no effect in our space-time. furthermore, it would not even be detectable.

As mentioned, we are exposed to singularity in each interval of Planck Time. This explains why the singularity is benign for us. The energy exposure is very brief.

If we confine our definition of physics to the limits of space-time, then we will need to renormalize our undesirable findings, frequently. Renormalization means using mechanisms to omit zero and infinities and other problematic findings which is not acceptable with our space-time knowledge and logic. However, the Big Bang theory, Dark Energy, and the nonzero Cosmological Constant are directing us to the beyond. There is more and more evidences now to force us to come out of our ken.

Pertaining to the sponge analogy, the Planck distance corresponds to the holes in the sponge. This implies that our universe is not continuous; rather, it has a discrete atomic structure. In the 1970's Jacob Bekenstein proved it by his discovery, the so-called "Bekenstein's Bound".[27] This law states that the amount of information within any horizon, forming a boundary, is finite and proportionate to the area of the horizon. If this is the case, the space inside the horizon cannot be continuous. It has to be finite and as a result has to be discrete.

Loop Quantum Gravity

Loop Quantum Gravity Theory is the next candidate in line for the Theory of Everything. The first being the Super String theory.

Lee Smolin, one of the main advocates of the Loop Quantum Gravity, claims that black holes thermodynamics, loop quantum gravity and string theory all agree that, on the Planck scale, space appears to be composed of fundamental discrete units. He adopts Roger Penrose's spin network as a model for fabric of the universe.

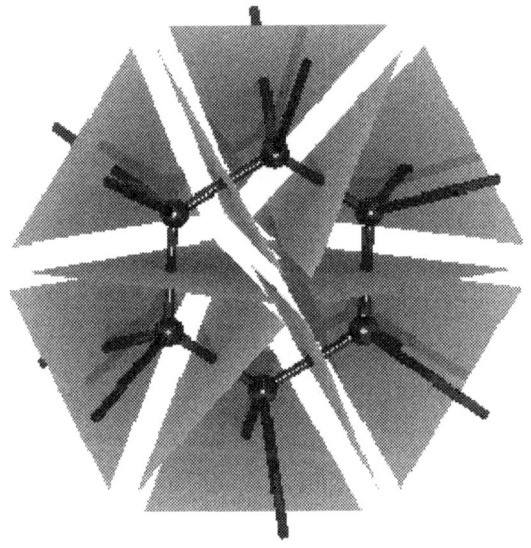

Spin Networks[28]
In this picture, each node represents a Planck space, which is separated from each other by the edges.

In the 1950s, Roger Penrose developed the concept of spin networks in an attempt to prove that physics terms are discrete rather than continuous quantities. It was originally developed to show that "everything was to be expressed in terms of *relation* between objects, and not between an object and some background space."[56]

However, if we look at the spin network in a large enough scale, it could be related to ordinary 3-dimensional Euclidean geometry.

The loop Quantum Gravity calculations also suggest that space is quantized. This theory postulates that space is created by its building blocks which are Planck distance cubic or $(10^{-33})^3$ or

91

10^{-99} cubic centimeter. Thus predicts that there are 10^{99} quanta of space in each cubic centimeter. Fortunately, this assumption may likely be testable in the near future.

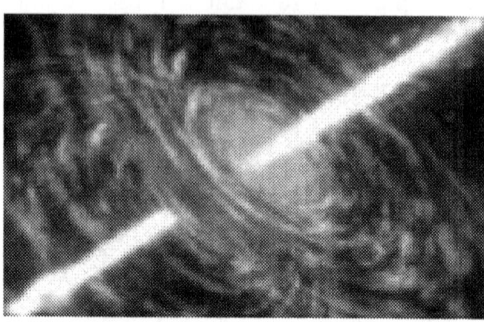

Examining the radiation from distant cosmic explosions called gamma ray bursts might provide a test whether fabric of space is discrete. If it is, radiation with different energy will arrive at earth at different times. The difference is minimal but for faraway bursts (3-4 billion light years away), it can be significant enough to be tested.

There is a modified version of Einstein's theory, by Lee Smolin and researchers from the Imperial College of London that accommodates high-energy photons traveling at different speeds. Our technology today is not precise enough to measure such a minute difference, but fortunately, the Glast Satellite, which is scheduled to be launched by NASA in near future, will have the required sensitivity for this experiment.

So, if the Glast Satellite shows that space is discrete, we can suppose each Planck volume (10^{-99} cubic centimeters) defines internal boundaries of space-time.

The Riddle of Infinities and Renormalization
In the introduction to this book I claimed that renormalizing of the theoretical physics mathematics is actually ignoring the road signs to explore reality. I further claimed that by normalization we create blockades for ourselves along the road. Steven Weinberg explains:

> "The standard model (of sub-atomic particles) is a quantum field theory of a special kind, one that is 'renormalizable' this term goes back to the 1940's when

physicists where learning how to use the first quantum field theories to calculate small shifts of atomic energy levels. They discovered that calculations using quantum field theory kept producing infinite quantities."[49]

The emergence of infinities in calculations was interpreted as flaws. It meant that the process was pushed beyond its limits of validity. This is done because the finite universe cannot include infinities. Since infinity is not physically possible, therefore the answer was interpreted as impossible and flaw.

Physicists consider infinities, which occur in ultra-short scales, a major problem. We can review the gravity field equation as an example:

$$F = G \, m_1 \, m_2/d^{\,2}$$

Where F is gravity force, m_1 and m_2 are masses of the two objects, G is Gravitational constant and d is distance between them. (As distances between the sun, earth, moon, etc...).[40]

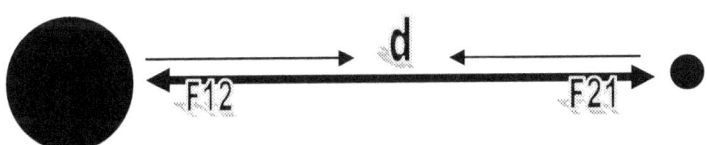

As d approaches to zero, gravity force increases and in zero distance (beyond Planck length) the force of gravity equals to infinity.

$$F = G \, m_1 \, m_2/d^{\,2}, \text{ if } d=0, \text{ then } g = G \, m_1 \, m_2/0 = \infty$$

This is also true for electromagnetic force that is inversely related to square of distance.

The act of ignoring and bypassing the infinity in calculations is called renormalization. Theoretical physicists use the Bekenstein Law (mentioned above) to get rid of infinities in the calculations.

If we confine ourselves to the physics which we believe rules only inside the space-time, then we have to frequently renormalize our findings. However, the Big Bang model, the proposed dark energy, and nonzero cosmological constant make us look beyond space-time. More and more evidence is there to force us to come out of our ken and believe that actual physics should extend beyond the familiar universe. If this is accepted, there would be no further need to renormalize.

Let us use common sense mathematics. Imagine that we have a bunch of pebbles. Now, we remove them from the scene one by one. At the end, there will not be any pebbles left. The reality is that there are no pebbles in the scene, but zero is still there. The scene still exists.

In conclusion, zero and infinity should also be considered a part of reality. So they cannot be ignored and avoided. Zeros and Infinities have profound effects on our world. We just need to give them an identity and insert them in our theories, not dismiss them as nothing or meaningless entities. The fifteenth century mathematicians had to expand the mathematical arena to include negative numbers. It took us until the nineteenth century to relate a physical meaning to them, namely the negative charges in electromagnetic. Just remember that there are also imaginary numbers, which we have not found an exact physical meaning for them yet. The actual physics extends to territories, which are vast, active and effective.

The presence of constants in our calculations indicates the unknown factors, which are affecting our universe. It is not humanlike to accept that these factors are out of our reach. We are hunters in the dark; and, so far, we have been finders.

Final Theory

Mark McCutcheon, in his book *The Final Theory,*[60] raised some very interesting questions. If gravity has been at work curving space and therefore pulling objects for billions of years (like the gravitational force between earth and moon which lasted at least four billion years), the energy has had to diminish as time goes by. If the force comes from the mass of the earth, we should have to see or foresee the end of it. Apparently, this force has been in effect for more than four billion years without any change. Where does this energy come from? This is obviously a breech of the conservation of energy law. McCutcheon answers this question by introducing the expanding world model. He rejects the notion of gravity and reasons that the attraction is actually because of the expanding planet, which swells out and reaches the free flowing objects. It seems that this model has serious flaws. A more logical answer is the aforementioned zero point energy. Precisely the energy from singularity that is infinite in this model. The mass of particles somehow introduces this energy to space-time, curving it according to the General Relativity.

With this definition of singularity, infinities in quantum scale are not problematic. It is defined and explained.

By assuming, the benign naked singularity as a medium that accommodates the core fabric of the universe, many paradoxes in physics will obtain a deterministic explanation. In this view our material world and ourselves are connected to singularity at all times and places. In this model, singularity is not 15 billion years away (birthrate of the universe). Singularity is here.

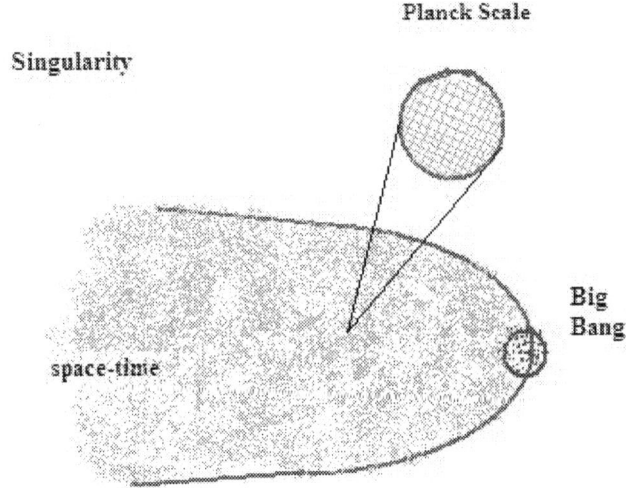

Singularity is here

Summary

Space in this model is a complex Minkowsky manifold. Which means it has three spatial dimensions and one time dimension? But each one of these dimensions are represented with a complex number (containing an imaginary element *i)*. As such it mimics twister theory, which is a Theory of Everything supported by Roger Penrose the famous British mathematician. But in my model the imaginary element is common between all four dimensions.

In this view the outer boundary of space-time and Planck distance and time are the edges where our universe meets singularity. As an analogy, we can imagine that our universe is dissolved and expanding in an infinite entity. I have also attributed the notions of zero and infinity to this entity. The concept of imaginary numbers is also accredited to physical activities related to this entity.

Holonomic Brain Theory

The Holographic Theory of Mind can provide explanations for two main puzzles, the nature of mind and non-locality. Therefore, it is important for us to investigate it.

Holography

In 1947, Dennis Gabor discovered the original optical holography. He showed that the information pattern of a three-dimensional (3-D) image could be encoded in a beam of light. Later on, the discovery of the laser helped put the idea into experiment.

"A hologram is a three- dimensional photograph made with the aid of a laser. To make a hologram, the object to be photographed is first bathed in the light of a laser beam. Then a second laser beam is bounced off the reflected light of the first and the resulting interference pattern (the area where the two laser beams commingle) is captured on film. When the film is developed, it looks like a meaningless swirl of light and dark lines. But as soon as the developed film is illuminated by another laser beam, a three-dimensional image of the original object appears."[58]

The feature of a holographic film which is of interest to us is the non-locality of its image. If we cut the film in smaller pieces every piece has the whole information of the original image so each illuminated piece still shows us the whole image.

Twentieth Century physicist David Bohm believed that the reason subatomic particles are entangled, even though they are far apart from each other, is because:

"At some deeper level of reality such particles are not individual entities, but are actually extensions of the same fundamental something."[58]

He considered this something a super-hologram that contains the information about past, present, and future and also includes the spatial data. In my model, I have assumed that this 'something' is the singularity.

According to Louis de Broglie the French Physicist and Nobel Prize winner, any particle or object has an associated wave motion. We also assume in the Particle-Wave Function chapter that every particle during its wave cycles enters and exits singularity. That is how interconnection and entanglement is achieved.

One can say that the image of singularity that I am trying to draw above is an extended version of holographic theory.

Fourier Transform

Let us look at the light and its data-transferring ability. When the sun-light reflects at a distant mountain, all the information is restored in a beam of light, which is heading towards us. To a certain extent, it does not matter how narrow you choose that beam of light, when we conjugate the information by using the lens of a camera, we still get the whole picture of the actual mountain. Moreover, depending to the strength of your lens, you could recover the information about surface texture even at microscopic scale of each point in the mountain. If we had a stronger device we could even extract the atomic or even subatomic structure information of every miniscule of the distant mountain. If you think about it, this is a lot of information for a tiny beam of light to carry.

The Bekenstein-bound, put a limit to amount of information that we can get from a screen with a limited area. The number of bits of information available will be less than one quarter of the area of the screen in Planck units. Nevertheless, it is still tremendous amount of information. Let us see how this happens? How do we recognize a spatial object, which is located say about 20 km away from us? You will say the light hit the mountain and part of it reflects and travels to our location. A part of the beam of light passes through our eye lenses and hit the retina. From there the action potential transfers the information to our brain and somehow our brain interprets it. In this way, we come to realize that there is a mountain twenty kilometer away. Let us investigate it further. Originally, the beam of sunlight was just carrying the information about surface of the sun. After it hits the distant mountain, it takes the bulk of information from our spatial object and includes it in the light wave. Underneath I use Dr. Jeff Prideaux's interesting description of holography.

"The act of converting spatial forms to frequency domain is determined by Fourier transform formula. The Fourier transform (and inverse Fourier transform) consists of convolution integrals, which mathematically smear or de-smear the information. For continuous functions, the Fourier transform and inverse Fourier transform are as follows (transformation between the time and frequency domain):

$$X(F) = \int_{-\infty}^{\infty} x(t)e^{-j2\pi Ft}dt \quad x(t) = \int_{-\infty}^{\infty} X(F)e^{j2\pi Ft}dF$$

The Fourier transform also has meaning between a spatial domain (for instance the position in two dimensional-space) and spatial frequency. Mathematically, the two-dimensional spatial Fourier transform is:

$$F(\alpha,\beta) = \int_{-\infty}^{\infty} \int_{-\infty}^{\infty} f(x,y)e^{-j2\pi(\alpha x + \beta y)} dx\,dy$$

And the inverse transform is:

$$f(x,y) = \int_{-\infty}^{\infty} \int_{-\infty}^{\infty} F(\alpha,\beta)e^{j2\pi(\alpha x + \beta y)} d\alpha\,d\beta$$

Where x and y are spatial coordinates and a and b are horizontal and vertical frequencies." [19]

When we put a lens and screen in front of our beam of light we change the frequency nature of information and convert it to a spatial image. Here we are doing the inverse Fourier transform.

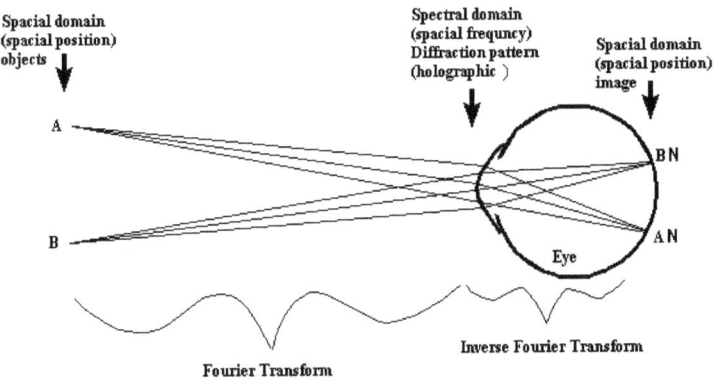

Diagram expressing the holographic nature of light incident on the surface of the lens of the eye[19]

Please note that the light by nature is electromagnetic energy not spatial. This is exactly the same concept that I am trying to convey about singularity. The proposed singularity contains information and can accommodate the information of four-dimension space just like the beam of light, which accommodate the information of our object, and its space.

Non-Locality

If we are at the foot of a mountain, and are trying to climb the mountain and reach the peak, it would need a lot of time and effort. That is how we actually notice the distance in human terms. If we use a helicopter, it takes less amount of time. Even if we go with the speed of light, although we reach the peak much faster, but it still takes few fractions of second to go from the foot of the mountain to the peak. However, there is no distance between the foot and the peak in our beam of light. One can choose smaller and smaller diameter beams and still get the information of far apart points in it (Note # 1). In a beam of light, all of the sudden there is no locality and everything will fall on top of each other. So we face non-locality in frequency domain. A lens helps us to diffract the information from the beam of light and the screen assists to extract the information. A lens and a screen apparatus perform the inverse Fourier transform and convert frequency information to spatial information, which we can interpret and understand.

101

Hologram just demonstrates the non-locality of information in spectral phase. In 1993 the famous Dutch theoretical physicist G.'t Hooft put forward a bold proposal. This proposal, which is known as the *Holographic Principle*, consists of two basic assertions:

"**Assertion 1:** The first assertion of the Holographic Principle is that all of the information contained in some region of space can be represented as a `Hologram' - a theory that `lives' on the boundary of that region. For example, if the region of space in question is a Coffee shop, then the holographic principle asserts that all of the physics, which takes place in the coffee shop, can be represented by a theory, which is defined on the walls of the Tearoom.

Assertion 2: The second assertion of the Holographic Principle is that the theory on the boundary of the region of space in question should contain at *most* one degree of freedom per Planck area."[18]

Before, I have assumed that the information in space-time, in it's entirely, is reflected and registered in singularity. To make it objective, Holographic theorists convert the whole ordeal to spatial form again but with one dimension less and present it to us. On the context of the proposed model, we can ignore the inverse Fourier transform and imagine that information remains in spectral state while in singularity. We do not have to pass to spatial phase (do not have to conjugate) and look at the shadow in the wall to realize that information is out there. Or we may do that for objectivity reasons, but at least we'd better appreciate and recognize the spectral state of the information. This is similar to mind function. According to holographic brain Theory, information remains at spectral form in the brain. That is what I am trying to convey about the singularity as well. The Holographic theorists stay in the boundary. I am not sure why we have to stop there. We have enough information to dare passing the border; at least our imagination should help us building theories and present them for speculation and investigation. Meantime if we establish a sound theory for mind function, we can utilize mind activities as analogy to explore beyond finite world. Holographic theory, says that all the information can be present in space with one-dimension less. Holographic theorist and M theorist (representing different Super String theories) found out that the answer of major paradoxes could not be found in our 4-dimension space-time. To find solutions they had to look out of our tangible space. My question is why did they have to travel to assumed spaces with different dimensions to create a basis to solve the paradoxes? Why couldn't we untie and free ourselves from space boundaries? We know from the Einstein's Special Theory of Relativity that time and space are not absolute.

At this point, let me add this beautiful piece from the University of Cambridge DAMTP web page.[18]

To them, I said,
the truth would be literally nothing
but the shadows of the images.
Plato, *The Republic (Book VII)*

Holography through the Ages

"Plato, the great Greek philosopher, wrote a series of 'Dialogues', which summarized many of the things, which he had learned from his teacher, who was the philosopher Socrates. One of the most famous of these Dialogues is the 'Allegory of the Cave.' In this allegory, people are chained in a cave so that they can only see the shadows, which are cast on the walls of the cave by a fire. To these people, the shadows represent the totality of their existence - it is impossible for them to imagine a reality, which consists of anything other than the fuzzy shadows on the wall. However, some prisoners may escape from the cave; they may go out into the light of the sun and behold true reality. When they try to go back into the cave and tell the other captives the truth, they are mocked as madmen. Of course, to Plato this story was just meant to symbolize mankind's struggle to reach enlightenment and understanding through reasoning and open-mindedness. We are *all* initially prisoners and the tangible world is our cave. Just as some prisoners may escape out into the sun, so may some people amass knowledge and ascend into the light of true reality. What is equally interesting is the *literal* interpretation of Plato's tale: The idea that reality could be represented completely as `shadows' on the walls."[18]

At the time of recovery of information we need a spatial arena to use as a scene to display the information.

Holonomic Brain

On the other hand, numerous studies in neuro-physiology suggest that memories in the brain are not stored in a specific location; rather, they are dispersed over the entire brain.

The conventional view is that the brain is a computational device. There is a growing body of literature, though, that shows there are severe limitations to computation (Penrose, 1994; Rosen, 1991; Kampis, 1991; Pattee, 1995). For instance, Dr. Jeff Paradeoux writes:

"Penrose uses a variation of the "halting problem" to show that the mind cannot be an algorithmic process. Rosen argues that computation (or simulation) is an inaccurate representation of the natural causes that are in place in nature. Kampis shows that the informational content of an algorithmic process is fixed at the beginning and no "new" information can be brought forward. Pattee argues that the complete separation of initial conditions and equations of motion necessary in a computation may only be a special case in nature. Pattee argues that systems that can make their own measuring devices can affect what they see and have 'semantic closure.'"[19]

As mentioned before, experiment shows that selective damage to certain area of brain tissue will not erase the specific related memories. It further suggests that memories are restored as frequency.

The experiment performed by Bernstein is worth mentioning. Here is a summary of his experiment and the follow up work by Karl Pribram, Professor Emeritus at Stanford University and his associate:

"Bernstein dressed people in black leotards and had them perform simple tasks such as running or hammering nails against a black background. The leotard had been decorated with white dots over each joint. Bernstein took cinematographic films of these activities. On his films he therefore had a record of the movements of the dots, which described a series of waveforms. When he analyzed the records according to a Fourier procedure he was able to accurately predict the next movement in the sequence. What we needed was direct proof that cells in the motor cortex were responsive to wave forms. So Amad Sharafat, an engineering student, and I devised an apparatus, which moved a cat's paw up and down at different frequencies. We recorded from motor cortical cells and found many that were tuned to the frequencies with which the paw was moved."[57]

He declares:

"What the data suggest is that there exists in the cortex, a multidimensional holographic-like process serving as an attractor or set point toward which muscular contractions operate to achieve a specified environmental result. The specification has to be based on prior experience (of the species or the individual) and stored in holographic-like form. Activation of the store involves patterns of muscular contractions (guided by basal ganglia, cerebellar, brain stem and spinal cord) whose sequential operations need only to satisfy the 'target' encoded in the image of achievement"[57]

If the muscle response has a wave pattern the message received has to have a wave pattern as well. May be it is allowable also to assume that the original messenger has also a wave-like (spectral) nature.

When, in 1960, Karl Pribram encountered the concept of holography, he used the concept to explain memory storage in the brain. After all, the capacity of the human brain to store and process information far exceeds the capacity of a spatially bound nervous system. He therefore concluded that the information must exist in a spectral form. He also considered the brain a hologram.

"It has been found that each of our senses is sensitive to a much broader range of frequencies than was previously suspected. Researchers have discovered, for instance, that our *visual systems are sensitive to sound frequencies*, that our sense of smell is in part dependent on what are now called 'osmic frequencies,' and that *even the cells in our bodies* are sensitive to a broad range of frequencies. Such findings suggest that it is only in the holographic domain of consciousness that such frequencies are sorted out and divided up into conventional perceptions."[58]

Using Holonomic Brain Theory to explain telepathy and other para-psychologic phenomena, Stanislav Grof a known educator and experimental psychiatrist and founder of Transpersonal Psychology says:

"If the mind is actually part of a continuum, a labyrinth that is connected not only to every other mind that exists or has existed, but to every atom, organism, and region in the vastness of space and time itself, the fact that it is able to occasionally make forays into the labyrinth and have transpersonal experiences no longer seems so strange."[58]

Mind as Hologram - Another Analogy

Earlier I mentioned that Holographic theory offers an explanation for the nature of mind. Now we can investigate it further. Experiments show that optical or other memories do not have to be stored in any specific location in brain. If any part of brain is experimentally damaged, still other live parts will show evidence of presence of stored memory. [Note 2] How are memories actually stored in the brain?

Karl Pribram says that both time and spectral information are simultaneously stored in the brain. He also draws attention to a limit with which both spectral and time values can be concurrently determined in any measurement (Pribram, 1991). The holonomic brain theory maintains that the brain is continuously engaged in correlation processes. This is how we make associations (how the senses are integrated). There is an obvious computational advantage for the brain storing sensory information (and perceptions) in the spectral (or holographic) domain as opposed to the brain cells directly storing individual features and characteristics. The holonomic brain theory claims that the act of "re-membering" or thinking is concurrent with the taking of the inverse of something like the Fourier transform. The action of the inverse transform (like in the laser shining on the optical hologram) allows us to re-experience to some degree a previous perception. This is what constitutes a memory. The Holonomic Brain theory (taking vision as an example) summarizes evidence that the image formed on the retina is transformed to a holographic (or spectral) domain. The information in this spectral "holographic" domain is distributed over an area of the brain (a certain collection of cells) by the polarization of the various synaptic junctions in the dendritic structures. At this point, there is no longer a localized image stored in the brain cells. Correlations and associations can then be achieved by other parts of the brain projecting to these same cells. Conscious awareness (and memory) is the byproduct of the transformation from the spectral holonomic domain back to the "image" domain. Possibly the most radical part of the holonomic theory is Pribram's claim that a "receiver" is not necessary to "view" the result of the transformation (from spectral holographic to "image"). He claims that the process of transformation is what we "experience". Memory is a form of re-experiencing or re-constructing the initial sensory sensation.[20]

(Conclusions made on the Holonomic Brain theory are based on Neuropsychological experiments. Please see the references below for detail of some of these experiments.)

Day Dreaming, Dream and Unconsciousness
With the holonomic brain theory the above notions can be explained as when the lens is partially or completely removed. Focus is lost, and we go to holistic phase or spectral domain.

Entropy

The second law of thermodynamics tells that the entropy always increases in any isolated system (see figure below). This simply means that if a system is left to itself, its energy distribution will move towards equilibrium or in other words it will move towards maximum disorder.

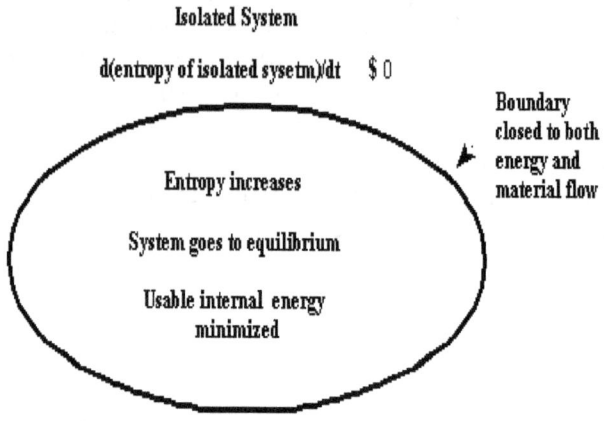

Entropy in a closed system[20]

If we take the space-time as an isolated system, then Second law of thermodynamics tells us that the universe has had maximal order and therefore minimum entropy in the beginning and is going towards maximum entropy and minimum internal organization as we go on.

At the surface, it seems that observation is pointing to the contrary. Reviewing the history of the universe, not only denies progress to maximal disorder, but it actually suggests that it is moving to obtain more complex and sophisticated structure as we go along. Universe progressed from creating sub-atomic particles to atoms of lightweight. Second and third generations of stars are creating heavier elements. From there simpler molecules are generated and they further developed themselves to more compound and complex organic molecules and their sophisticated functions.

But, in reality, many of these phenomena are part of a bigger process. During the main process the amount of disorganization and heat release increases and surpasses the formation portion of the process. Therefore, we cannot consider the formation part as an isolated system. We have to look at the whole system where the disorder prevails.

Briefly, it means that an isolated system can contain a subsystem that is open to energy flow from the main system (see figure below). As such, the whole combined isolated system still obeys the second law of thermodynamics, but it is possible that the subsystem can experience a decrease in entropy at the expense of its environment (the main system).

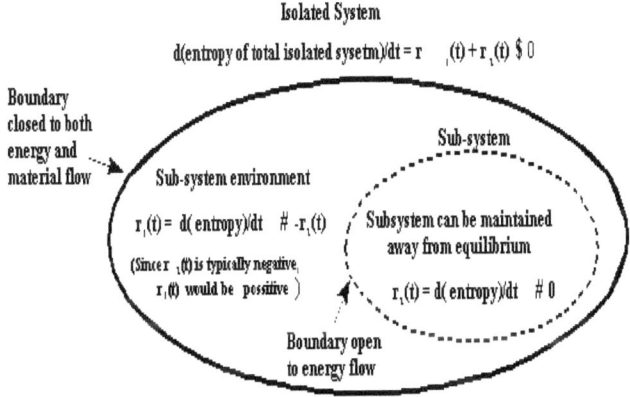

Entropy in a system containing a subsystem[20]

Sun a Helium Factory

We know that the sun is actually a helium factory. Inside the core of the sun, at the temperature of 15,000,000° C, four hydrogen nuclei fuse together to form one helium nucleus. This suggests that we are moving toward a constructive procedure. However, helium nucleus is about .7 percent less massive than four protons. The difference in mass is released as energy that leaves the sun and enters the surroundings. Therefore, the constructive procedure has not entirely happened inside an isolated system. The dissipated light and heat in turn give us warmth and light on planet earth, and is the main source of all the creativity on earth. We all see the constructive effect of the sun energy. However, ultimately this energy leaves the solar system and dissipates throughout the universe. Therefore although some constructive phenomena have been gained through the process, this gain is at the expense of increased entropy and disorder throughout the universe.

But there are other events that cast shadows over the law of entropy increase. The heavier atoms are manufactured inside stars through the fusion of lighter atoms. Formation of atoms to iron is accompanied by releasing energy, which is in line with the second law of thermodynamics. All the elements heavier than iron are created in huge supernova explosions.

Many anabolic chemical reactions are so-called exothermic, which means the reaction is accompanied by releasing energy.
We can write the formation of water as:

$$1 H_2 + \frac{1}{2} O_2 \rightarrow 1 H_2O + 68.3 \text{ kcal}$$

This is in line with the Second Law because it releases energy. However, there are constructive reactions that are endothermic, which means they absorb energy.

$$\frac{1}{2} N_2 + 3/2 F_2 \rightarrow NF_3 - 27.2 \text{ kcal}$$

However, in these occasions we can claim that the endothermic part is just a portion of a main reaction. The total energy released during the main reaction is more than the energy absorption for the endothermic portion of it.

Yet there are exceptions. The formation of the early universe from a burst of energy is an example of such an exception where this explanation is not valid. How did the burst of energy create the homogenous low entropy baby universe?

Black Holes are another exception. The singularity inside them has the ultimate entropy and disordered state. Nevertheless, as they vaporize, they vanish and a state of less entropy substitutes their existence.

Positive cosmological constants point to the presence of dark energy as a causative factor. The notion of dark energy also is an exception to the law. Francisco Di Biase and Mário Sérgio F. Rocha from Dept. of Neurosurgery/Neurology and Computed Tomography Santa Casa Hospital write:

"The entropy increase in the "sub-system environment" is guaranteed (by the second law) to more than offset the entropy decrease in the subsystem. Also note that the sub-system can only be maintained away from equilibrium as long as there is usable energy in its environment."[20]

The above model explains the chain phenomena in our space-time universe. Although the main order is entropy, organization and differentiation continues in the subsystems inside space-time. Therefore, if a low entropy state is created in a zone, we have to take that zone as a subsystem and look for the surroundings and include the surroundings as part of the system. Then we can expect that the entropy will increase in the entire zone. If a low entropy state is created at the time of the Big Bang, we have to look outside of the newborn universe for a high entropy zone. This would be the singularity. In fact, singularity by definition has maximum entropy.

In my model, the singularity/space-time is the main system where the space-time is just a subsystem. Many quantum mechanical phenomena suggest the violation of conservation of energy and matter (the first law of thermodynamics). That is, if we take space-time as a closed system. However, if we take it as a subsystem of a main system, then all of those violations will obtain explanations. Therefore, this view bypasses the problems with the first law of thermodynamics.

Therefore, in this view, the universe can accept structure and organization at the expense of entropy in singularity. Since we assumed the energy of singularity to be infinite, the space-time can continue building its internal structure forever. In this model, the entropy inside the space-time can decrease wherever needed and order prevails. For this model to be acceptable, we have to assume that exchange of energy is possible in the boundaries of our universe. On the other hand, our world is not a place for random and unrelated structures. It seems that our world is goal oriented and follows a pattern of self-organization. As De Biase states, it continuously creates and recreates itself and explores the possibilities of new existence. Upon destruction of first generation stars and from their dust evolves second generation of stars, which are the source for heavier and more complex elements. Third and fourth generation stars follow the mission and create even more complex atoms.

In addition, the changes in far apart locations in universe are similar and follow the same pattern. We may assume that because the initial conditions have been the same and the laws of physics are similar, the changes in distant locations are alike. However, the possibilities of changes, especially at subatomic level are endless. If the world follows the same pattern in distant and far apart locations, we may conclude that, world is interconnected and goal oriented. For the world to be interconnected, it requires a non-local media. We have taken the proposed singularity as being this media.

Going back to Holonomic Brain Theory and its interpretation of brain function, we again encounter with similarities between singularity and mind.

"There is a special class of such subsystems (as described above) where the subsystem's organization comes exclusively from processes that occur within the sub-system's boundaries. This class of subsystems was labeled "dissipative structures" by Prigogine, 1984 (who won the Nobel Prize for his work)."[19]

To explain how the holonomic brain functions, Pribram suggests "dissipative structure" as a model. One way of modeling a structure that goes to equilibrium is to minimize a mathematical expression for the internal energy (which is the same as maximizing an expression for entropy). Interestingly those factors have been present in every miniscule of the universe, thus, we are observing a logical pattern which is called the least action principle. The principle of least action defines the action S for motion along a world-line between two fixed events. This would not be appropriate, though, for a "dissipative structure" since it is not going towards equilibrium. "Dissipative structures" self-organize around a different "least action principle". In the holonomic brain theory, Pribram states that in the brain, the entropy being minimized (which maximizes the amount of information possible to store) as the "least action principle". Thus, the system (the brain) self-organizes such that more and more information can be stored.

We can use the same subsystem model for the universe and proposed singularity.

Singularity

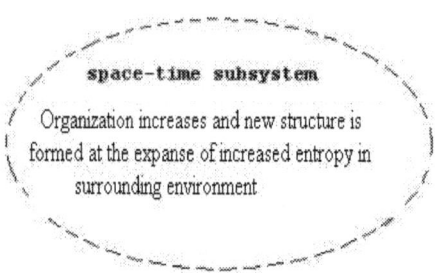

space-time subsystem

Organization increases and new structure is
formed at the expanse of increased entropy in
surrounding environment

Dissipative Subsystem Model

Perideoux writes:
"In Hopfield networks and the Boltzmann engine (which are computer models of neural processing), computations proceed in terms of attaining energy minima. In the holonomic brain theory, computations proceed in terms of attaining a minimum amount of entropy and therefore a maximum amount of information. In the Boltzmann formulation the principle of least action leads to a space-time equilibrium state of least energy. In the Holonomic Brain theory, Pribram describes the principle of least action as leading to maximizing the amount of information (minimizing the entropy). Independently, (in unrelated work) Schneider and Kay (1994) have proposed a variation on the second law of thermodynamics, (The entropy of an isolated system which is not at equilibrium will tend to increase over time, approaching a maximum value.) which may be applicable to Pribram's Holonomic theory."[19]

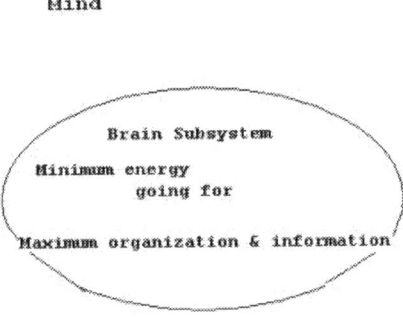

Mind

Brain Subsystem
Minimum energy
going for
Maximum organization & information

Dissipative Subsystem Model

Universal Mind and Individual Brain

Wheeler (1990) and Chalmers (1995) realized how important the information is in such context. Chalmers, by stating that information must be considered as an essential property of reality as matter and energy, and that "conscious experience be considered a fundamental feature, irreducible to anything more basic". Wheeler, with its famous "*the it from bit*" concept that allows us to unite information theory to consciousness and physics writes:

"...Every it – every particle, every field of force, even the space-time continuum itself – derives its function, its very existence, entirely - even if in some contexts, indirectly - from the apparatus-elicited answers to yes-or-no questions, binary choices, bits. Norbert Wiener put this identity on the very conceptual basis of cybernetics stating that 'information represents negative entropy', and prophetically emphasizing that 'information is information, not matter or energy.' Consciousness is conceived as a non-local flow of meaningful quantum-informational activity, interacting actively with each part of the universe through the holo-movement."[20]

Conclusion

So according to Dr. Pribram, consciousness does not need a part of nervous system to physically accommodate it. Although he

115

believes "the transformation which is a spread function spreads a pattern within the confines of a sensory receptive field (brain)."

Mind you that one of the major hypothesis in this model is that mind is an extension of singularity. Holonomic Brain theory asserts that the consciousness has a spectral nature. According to Dr. Pribram radioastronomy patches of holograms can be seamed together to spread the pattern to a larger territory retransforming it into space-time. Stanislav Grof suggests that mind is interrelated with other minds and is part of a continuum.

Here I propose that consciousness is out there without any physical base. Just like information, which is out there? We use lens and screen to display the information embedded in a beam of light. Similarly, we need a spatial scene and physical tools to illustrate the consciousness. So I view the mind, as a separate entity from so-called physical brain, similar to Plato's view.

The Holonomic Brain theory suggests that conscious awareness is a state of wave function and not a material based entity. In this model, I also postulate that information is present in no dimension zone, in an entity out of spatial domain. If we know as a fact that space can be expanded and contracted, we also have to accept that space can cease to exist, as it would happen in the pre-Big Bang era. I freed myself from ties. I passed the boundaries. I jumped off the cliff. And guess what? It was not dangerous or scary at all. There was not just darkness. A new tangible and deterministic world exists out there. We can take the risk and receive the rewards. Now it seems to me that the major paradoxes possess explanations if we leave our ken. We will look at those paradoxes later together.

Notes

1) The above statement is not exact. In distances smaller than the light wavelength, the information courier system falls apart. Please note that here the hologram is used as an analogy to demonstrate how non-local information can convert to a local image. A justification about dissolution of information in wavelength distances can be derived from the *particle-wave function* section in this model.

2) "A series of experiments were performed in both cats and monkeys (De Valois et. al., 1979) to see if the cortical cells responded to differences in the Fourier spectrums. The results showed that, the visual cortical cells respond to the Fourier fundamentals, not acting as an edge detector. In (De Valois et. al., 1979) experiment visual cortical cell respond to the angular location of the Fourier fundamentals and not to the edge of the squares (or grating) seen in the untransformed pattern. It also showed that the visual cortical cell was responding to the Fourier fundamental and not the edges (or distance between the edges) of the visual stimuli." Please check Reference 19 for detail.

The Big Bang Theory

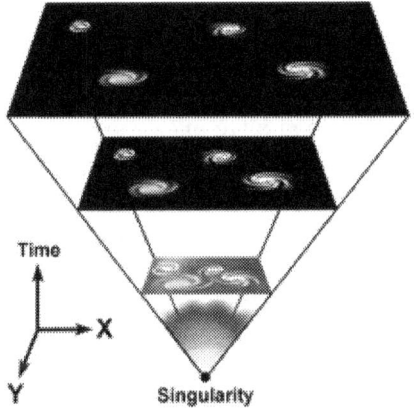

The Big Bang Theory is the prime theory to describe the beginning of the universe. George Gamov first suggested the concept in 1948. The theory explains how the universe emerged from a tremendously dense and hot point about 13.7 billion years ago. Precise prediction of background radiation temperature, even before its actual discovery, and recent findings, which suggest that our universe was much hotter in the younger ages, supports the Big Bang Theory. However, physicists have mainly avoided the conundrums of an infinitely compressed zero-sized starting point for Big Bang. The problem arises if we go to diameters smaller than Planck length. We cannot find any meaningful smaller size in our universe. The only way that we can pass beyond this size is, if we exit our materialistic world and think singularity. Our objectivity based scientific laws do not govern in this arena. Any attempt to carry over these laws is doom to assumptions, uncertainty and confusion.

Origin of Matter

Although there are strong evidences to support the main theory, there are different opinions about how the Big Bang event actually happened and progressed.

119

In Gasperini and Veneziano's pre-Big Bang scenario, the universe started out as cold and essentially infinite in spatial extent. In addition, they are suggesting that Big Bang is not the initial event in creation of our universe; rather, it is a few steps further.

Guth Inflationary Theory

The inflationary theory is the most popular and accepted version of the Big Bang. In the inflationary model, the initial stage contained a very rapid and huge expansion of the space. This just took a fraction of a second. Then the expansion slowed down for about seven billion years. This period is called deceleration stage where the rate of expansion had been slowed down by the gravity of the matter inside the universe. Then as space got bigger and thinner, the gravity got weaker. As a result, the expansion of the universe has been accelerating for the past seven billion years since.

Cosmologists consider the cosmological constant the factor behind the accelerating force. This constant is supposed to exert repulsive force to the space-time. It is believed that dark energy with its negative pressure is the source for the cosmological constant. For the expansion to accelerate, we need constant production of dark energy. Where does this energy come from?

Continuity Equation Principle

Continuity Equation Principle ^(Note 1) requires that the density of matter in a region of space stays the same. In addition, the density of the universe is currently close to the critical density. Which means the gravitational force from the matter/energy density inside the universe can counteract the expansion force and keep the universe almost flat (please see the Flatness Problem chapter). Chances are that the universe density has always been close to critical density.

The dark energy contributes to the density of the region. In fact, seventy percent of the density in the universe is attributed to dark energy. However, since the universe is constantly expanding, we would need constant matter creation for the continuity equation to hold. Where does the matter come from?

What is the origin of the additional matter or dark energy being continuously created? One is led to look at other possible scenarios for matter creation besides the initial Big Bang. We need another source readily accessible by any inch of space to supply the matter needed. Steven Hawking presents the lead.

"The uncertainty principle of quantum theory means that fields are always fluctuating up and down even in apparently empty space, and have an energy density that is infinite."[6]

Please note that infinite energy cannot belong to a finite world. He further continues:

"The universe may contain what is called Vacuum energy that is present even in apparently empty space...Vacuum energy causes the expansion (of the universe) to accelerate"[6]

Besides, infinite energy defies the conservation of matter and energy law (first law of thermodynamics). Many quantum mechanical and astrophysical observations put the conservation law under question. Positive cosmological constant may indicate that matter and or energy constantly penetrate inside the space-time. If we assume that dark energy can leak in from Planck's pores throughout our universe, the result would create the exponentially expanding universe.

On the other hand, expansion of the universe requires either the space to stretch or being built from inside. If we believe that space is a continuum, then it has to stretch to provide expansion. However if we take space as a discrete entity, then the building blocks have to be built from within. In this model, space is discrete. Therefore, we have to look for mechanisms that can create space blocks.

Steady State Theory

Steady-State Theory of Bondi and Gold suggests that, as the universe expands; new particles of matter are continuously created to fill up the gap so that the average density of the matter in the universe remains unchanged. Hoyle suggested "creation field "for creating matter from energy. He proposed presence of negative energy in his creation field to compensate for breaking the conservation of energy law. Dark energy can be the source for space unit creation. Then again, this challenges the conservation law.

If we are so keen to conservation law and would like to keep it anyway, then we have to expand it to contain a source of energy beyond the space-time (singularity in this model). But then if singularity has infinite amount of energy the law looses its meaning. Alternatively, we can leave conservation law for domestic economy and take space-time and singularity as trade partners. Conservation law works very nicely in macrocosm. We can leave the boundaries energy transaction to microcosm and quantum mechanical arena. This is the arena where we encounter violation of the law the most. Roger Penrose in *The Large, The Small and Human Mind* writes:

"...All of the other schemes for quantum state reduction, which attempt to solve quantum measurement problem by introducing some new physical phenomena, run into problem with conservation of energy. You find that the normal rules of energy conservation tend to be violated. Maybe this is indeed the case."[5]

Dark Energy

Interpreting the conservation law in this manner will free our imagination, and we can create theories to explain some unexplained findings in astrophysics. The big challenge today is finding a source for dark energy. The presented model offers a solution for the origin of dark energy.

The general acceptance of The Big Bang Theory by cosmologists, does not exclude the possibility of steady mater creation as a contributing factor in formation and expansion of the universe. Of course, $E = mc^2$ gives the formula for this energy to create matter. Isn't it true that continuity equation [Note 1] should hold everywhere in space?

Expanding universe without matter creation is against the above equation.

If we try to uphold the conservation of energy law, then we have to assume the presence of negative energy in vacuum to compensate for permeation of energy and accelerating expansion of universe. With the above definition, the law is not broken. But if we extend the law to an arena beyond the space-time or question its universality, then we do not have to be worried about hypothesized negative energy. Then we can explore the possibility that zero point energy fluctuation in intergalactic space, can flourish and enter and provide the dark energy for the positive cosmological constant and acceleration of expansion.

Shape of the Universe[(Note 3)]

Steady matter creation also suggests a solution for *Flatness Problem.*[(Note 2)] What does flatness problem mean? If the density of our universe is greater than the critical density, in another word, If the gravitational force created by density of matter in the universe is greater than the expansion force, our universe is 'closed,' It means our universe will eventually stop expanding and start contracting. If the density of the universe is equal to the critical density, then we live in a 'flat' universe. Lastly, if the density of the mater in the universe is less than the critical density, then the universe is 'open.' In an open universe, the expansion continues forever. Currently, the best-known value for the critical density is about $1*10^{-29}$ grams per cubic centimeter. Recent measurements indicate that the actual density of our universe is very close to the critical density. Although, matter density of the universe is so close to unstable critical value between perpetual expansion and re-collapse into a big crunch, most recent researches suggest that the rate of expansion is increasing (1998 Perlmutter et al).

Steady Matter Creation

The proposed steady matter creation is needed to compensate for the ever-increasing expansion of the universe. It is needed to keep the actual density of the matter in the universe close to critical density. To accept the steady matter creation we do not have to deny Big Bang model. Matter creation in the Big Bang moment and matter creation in vacuum can follow the same principle. Except that they are working in two different scales, big and small.

New stars are formed through the nebula activity in different parts of the Universe. Old star's dust is the building blocks for new star formation. Is there a chance that newly formed matter is also contributing to star formation or dark matter lenses? [(Note 3)]

Notes

1) In electromagnetic theory, the **continuity equation** is derived from two of Maxwell's equations. It states that the divergence of the current density is equal to the negative rate of change of the charge density.

2) Check the Flatness Problem chapter.

3) Visible matter is a small portion of the matter present in the universe. In our Galaxy, the Milky Way, just about 10 percent of the matter is visible. It is suggested that the remaining 90 percent are dark matter which do not radiate so they are not visible.

Visible and dark matter creates only 30 percent of the critical density. The other seventy percent is believed to be in the form of dark energy.

Boundaries and Evidences

The Big Bang Theory implies that the space-time is not stretched indefinitely. The universe has boundaries. It has a periphery that is marked by the outer boundary of the expanding universe. In the present model, we picture the universe as a sponge. Therefore, universe also has internal boundaries or interfaces with the medium entity. Hereafter, I will call the inside boundaries *the Planck pores*. Furthermore I have assumed consciousness another boundary where the space-time intermingles with the singularity.

Many physicists like James Hartle and Stephen hawking believe that the paradoxes in theoretical physics can only be solved by specifying the boundary conditions of the universe. Here we explore the boundaries and try to look for evidences according to the current knowledge.

Before the brightness of day or the darkness of night we get the grays of dawn and dusk. Likewise the structure of the space-time pale whenever we get close to the interface between space-time and the proposed singularity namely the sub-atomic arena, mind and The Big Bang moment. In this chapter, I will show how the elements of space-time get pale as we approach the grays of dawn and set of our universe. That is when we get close to the boundaries.

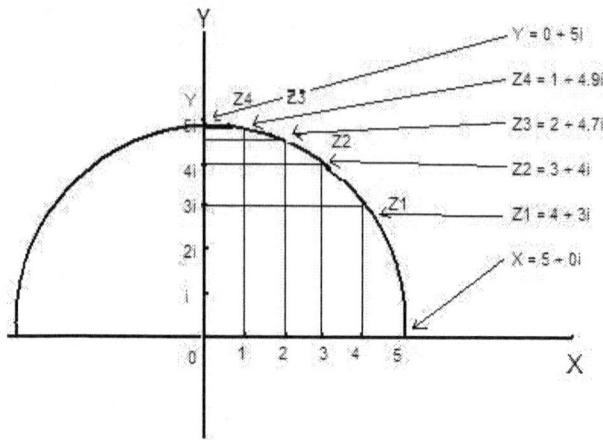

Gradual disappearance of space-time value and emergence of imaginary Value

Argand Diagram for Boundaries

Complex Numbers and our interpretation of them are explained in the first and the eight chapters (Complex Numbers and Wave Function chapters).

The diagram above shows how elements of space-time fade away as we approach the boundaries of universe and approximate the purely imaginary domain. This process is explored further in the lines below.

Matter

By definition matter is an entity, which can occupy a definite location in space. If there is no space, matter looses its definition. As it is shown underneath, matter gradually looses its entity when we get close to the boundaries of space. Underneath, we review matter in different arena.

126

A) Microcosm

As we get down to particle physics the concept of matter gets more subtle and pale.

The photon is mass-less and even its existence is under question. In 1969, Lamb and Scully showed that one could account for the photoelectric effect without using the concept of photon as a minimum packet of light energy. They were able to introduce an entirely different theory of the photoelectric effect, one that did not invoke the concept of light's particle nature. They concluded that, photoelectric effect does not prove that photon exists. In addition,,

George Greenstein writes;

"[In 1956] The Hansbury-Brown and Twiss experiment failed to demonstrate the existence of photons and the indivisibility of weak light. It actually showed that light seemed to travel through space "bunched up". One can divide the bunch in half, and the two half bunches arrive at the different photo detectors at the same time. These result startled the physics community and launched an entirely a new discipline, the explicit study of quantum nature of light."[11]

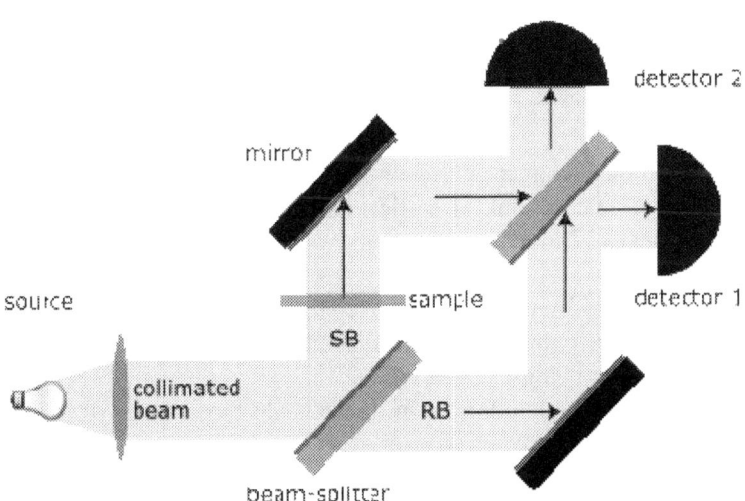

Mach-Zehnder Interferometer

Later on, the same experiment was repeated by laser, which still did not support the particle nature of light. In 1986, in the

127

Grangier, Roger, and Aspect experiment, the non-divisibility of a light unit was shown as evidence of the presence of photons. They used a well-collimated stream of calcium atoms. In their next experiment, they allowed the photon to pass through Mach-Zehnder interferometer. They obtained an interference pattern as a path length traveled by light in one arm of the interferometer was increased relative to the other. So, light divided and passed thorough both ways. Again, the result pales the concept of the photon. Greenstein and Zajong conclude:

"It is ironic that Albert Einstein, arguably the greatest physicist since Newton, received the Nobel Prize for work that subsequently turned out to be flawed. And it is doubly ironic that this work, which was instrumental in placing before us the concept of wave particle duality, turned out to be correct even though flawed…The central lesson of the story we have recounted…Is that the concept of the photon is far more subtle that has been previously thought."[11]

Wave particle duality also extends to the atom itself. Many experiments are performed passing one atom at a time through a double slit apparatus like the Mach-Zehnder interferometer shown on the previous page. Even one atom created an interference pattern. One expects to have two particles to create interference. How can just one atom demonstrate the performance expected from two atoms? Here, the solidity of individual atoms goes under question.

Erwin Schrodinger goes even further to claim:

"One can think of particles as more or less temporary entities within the wave field whose form and general behavior are nevertheless so clearly and sharply determined by the laws of waves that many processes take place as if these temporary entities were substantial permanent beings."[12]

So we can conclude that in micro-scale the solidity of matter gets pale.

To solve the mysteries of nature, we must raise odd questions and move toward bold ideas. Here is another speculation:

Standard Model of Particle Physics

The standard model of particle physics, which deals with and categorizes the subatomic particles, divides the known particles to two different groups. Fermions (spin-1/2) and gauge bosons (spin-1). All sixteen particles in the mentioned categories are already discovered and observed in high-energy colliders. Interestingly, all of these particles are mass-less. This is obviously in contradiction with nature because the particles and their products (atoms) have clearly possess mass. Therefore, we had to hypothesize and introduce another particle, the Higgs boson. This particle is supposedly responsible for the Higgs mechanism by which all other particles acquire mass. However, the Higgs boson has yet to be discovered.

We try to relate this particle or some other physical phenomena to the process of naturalization (giving mass to the sub-atomic particles). Is there a chance that at the gray area of smaller scale and beyond, mass does not exists? Is there a chance that mass is the kinetic energy of particles as they travel along their waves? Is there a chance that gravity is the byproduct of this wave motion of objects? We will discuss this issue in Mass and Gravity chapter.

Do we have to hypothesize quanta for gravity? Can this questioning open the road to explain the paradoxes between quantum mechanics and general relativity? This is just a speculation. But we mustn't leave any road unexplored.

It is general understanding that the principles of matter entity in macrocosm are gradually being violated as we approach the particle physics. In that level particles demonstrate their dual character of wave-particle. Even the identity and sameness of specific particles undergoes questioning. We cannot identify different particles from each other. At particle level electrons are the same and we cannot differentiate them from one another.

■■

B) Matter, Big Bang and Space-Time Singularities

According to the popular Big Bang Theory, matter gradually formed and appeared at the end of the initial rapid expansion. The beginning of time is stated at 10^{-43} sec. having a temperature of 10^{19} Gev. After the cosmic inflation, vacuum energy transforms itself into an equal number of particles and anti-particles of matter. Prior to the electro-weak era, small excess of quarks and electrons appeared over anti-quarks and anti-electrons. At 10^{-10} sec, force and matter fields differentiated. At 10^{-4} sec. Quarks, joined and formed proton and neutron. At 100 sec. the nuclei of light elements began to form. It took 10,000 years before transition occurred from the domination of the radiation to that of matter.[13]

The followings are the fundamental forces of nature and their relative magnitudes:

Strong nuclear force	10^{40}
Electromagnetic force	10^{38}
Weak nuclear force	10^{15}
Gravity	10^{0}

What is more, when the universe was 10^{-39} second old, strong, weak and electromagnetic forces were united. With the same token, while the above forces in larger scales differ greatly in magnitude (see the table above) as we examine them in distances about 10^{-29} centimeter (ten thousand larger than Planck distance) the three non-gravitational forces appear to become equal. It seems that in smaller scales the structural elements of space-time get simpler and melt away.

Therefore, at the beginning we don't have any trace of mass; Matter only appears as the universe develops.

C) Matter and Mind

We have considered the mind function as an analogy for the nature and function of the proposed singularity. Let us now use the same analogy to develop the concept further.

In the mind chapter, while talking about dreaming, we have seen that the dream house was an intangible mass-less image created within our consciousness domain. The actual house is absent in the realm of the mind.

What did happen when we stopped dreaming and made the decision to go out and by the lottery ticket? An intention originated in our mind-domain initiated a series of incidences in space-time. We had departed from our dream world and tried to materialize our dream. So with the postulation of mind as a being out of space-time universe, we see that matter fades away or flourish in interactions between mind and space-time universe.

Space
Black Hole

By definition space and time gets distorted and twist as it passes the event horizon of a black hole. They will be swallowed and disappear at the proximity of the singularity located at the center of each black hole. The image below shows how space comes to an end inside a black hole. Here we witness the dusk of space as we leave the space-time.

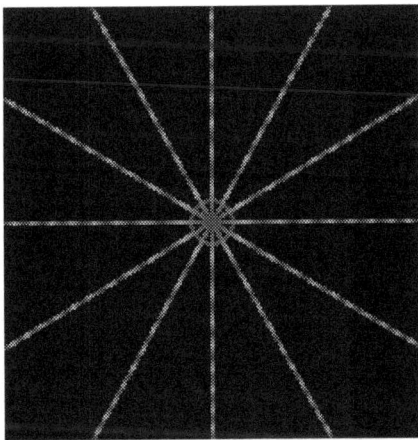

Space and time swallowed by singularity

Andrew Hamilton from University of Colorado describes how space and time is swallowed by a black hole:

"Free-fall coordinates reveal that the Schwarzschild geometry looks like ordinary flat space, with the distinctive feature that space itself is flowing radially inwards at the Newtonian escape velocity, $V = (2\,G\,M\,/\,r)^{1/2}$.

The in fall velocity V passes the speed of light c at the event horizon. Picture space as flowing like a river into the black hole. Imagine light rays, photons, as canoes paddling fiercely in the current. Outside the horizon, photon-canoes paddling upstream can make way against the flow. But inside the horizon, the space river is flowing inward so fast that it beats all canoes, carrying them inevitably towards their ultimate fate, the central singularity. Does the notion that space inside the horizon of a black hole falls faster than the speed of light violate Einstein's law that nothing can move faster than light? No. Einstein's law applies to the velocity of objects moving in space-time as measured with respect to locally inertial frames. Here it is space itself that is moving.."[16]

Space Phase Transition

The speculations about phase transition of space as it approaches infinite energy of proposed singularity are in line with Chapline presentation in Texas Conference on Relativisitc Astrophysics; Stanford, California, 12/12-17/04.

George Chapline from Lawrence Livermore, National Laboratory in California In his recent (Nature, March 2005) article "Black holes 'do not exist'" mentioned that collapse of big stars creates a zone which differs from ordinary space-time and contains much larger vacuum energy. He calls this zone dark energy star that is different from condensed mass zero-point singularity of a black hole. The surface of such a dark energy star:

"Corresponds to a quantum critical surface for space-time. The behavior of matter approaching such a quantum critical surface can be surmised from the behavior in the laboratory of real materials near to a quantum critical point. One prediction is that nucleons will decay upon hitting the surface of massive compact objects."[38]

This strange behavior, he says, is the signature of a 'quantum phase transition' of space-time. However, even he confirms the phase transition of space and time in the proximity of the vacuum.

Periphery of the Universe

We do not have much evidence coming from outer peripheries of the universe, for we do not have access to it. But if we believe that our universe is an expanding system, then one expects to see the signs of the outside entity at the outer boundaries of the premature space. Outside layer of the space that is under construction.

Space and Microcosm

Space gets pale in smaller scales. The Heisenberg Uncertainty Principle also suggests that the uncertainty of the particle's location in ultra small distances is fundamental and does not happen because we do not have sensitive enough instruments or because of errors in measurement, but because in subatomic scale particles do not possess locality. In these ultra small scales, location and mass pales.

Tonomura's Double Slit experiment (which is described in Quantum Mechanics Chapter) tells us that in ultra small scale, particles do not even have a specific trajectory. The experiment puts the presence of space and location in smaller scales under question.

Time

Microcosm

Time-energy uncertainty is an extension of the Heisenberg Uncertainty Principle. The uncertainty principle governs at smaller scale, therefore,

$$\Delta E \text{ (energy)} * \Delta T(\text{time}) \geq h(\text{Planck Constant})/2\pi$$

The above principle is also fundamental. The problem is not because our measurement apparatus is not sensitive enough. It happens because in smaller scales the notion of time becomes scrambled and it is interchangeable with energy. In smaller scales even long before we come to Planck Time, the notion of time gets blurry and vague.

133

Compton Scattering[23]

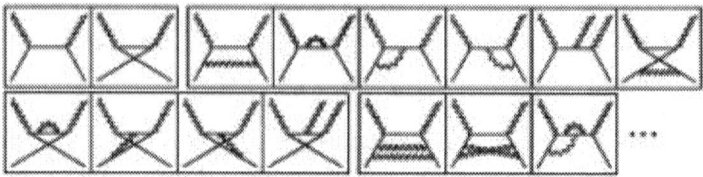

Compton Scattering refers to the way that a photon strikes and scatters from an electron. The diagrams above show some of many different Feynman Diagrams representing possible scenarios for meeting a photon and an electron. According to Richard Feynman all of these scenarios are simultaneously happening.

In some of the Feynman diagrams, we will see the distortion of time at the infinitesimal instant of the collision. Even in the second diagram, time reverses; this means that the particles scatter from each other before they even collide. The diagrams show how the sequence of events in a microcosm gets mixed up.

What is the time scale for Compton Scattering? If the energy of the incoming photon were 100 keV, then the intermediate state lasts for 6.6×10^{-21}s.[11]

This is still many times longer than the Planck time of 10^{-43}s. In comparison to Planck time, it is a very long dusk, like a fall sunset in the North Pole. In fact, Feynman Diagrams of Compton scattering suggests infinite expressions representing infinite ways that an electron and a photon can interact. Each of these possible ways is happening in a very short time. Most of these different possibilities cannot happen in our space-time. Different reasons such as the Conservation of Energy Law inside space-time, will not allow them to happen. We need a domain with an infinite amount of energy, which at the same time is not time bounded for these infinite possibilities to take place.

Time and Macrocosm

In the large scale, we are looking at the light cone, which initiates at the Big Bang and extends to this moment. So we may claim that the outer boundary of time is now, approximately 14 billion years away from the Big Bang moment. Beyond this instant, time has not reached yet. What we can foresee beyond this moment are potentials. The possibilities are purely determined by the existing data and available energy, the two elements that we have assumed for the singularity. Can we conclude that in the outer limit of time we face singularity in the large scale?

In addition, according to Einstein's Special Theory of Relativity, as an object accelerate from lower speeds to relativistic speeds (speeds approaching the speed of light) time slows down. Once in proximity of this speed, time will dilate and eventually disappear. Brian Greene writes:

"The maximum speed through space occurs if all of an object's motion through time is diverted to motion through space...but having used up all of its motion through time, this is the fastest speed through space that one object - any object - can possibly achieve."[1]

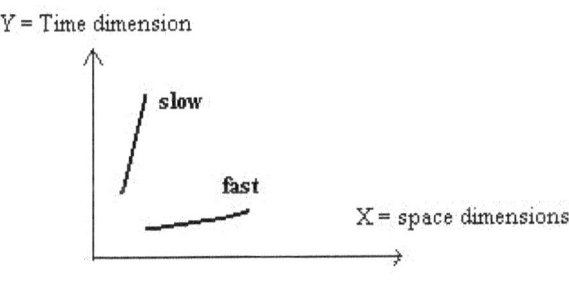

Speed
Movement of an object along time and space dimensions

Therefore, if we are traveling with full speed (speed of light) in space, we are not traveling in the time dimension at all. So, we may conclude that in outer space boundaries (borderline with singularity) there is no time.

Energy

According to our original assumption, energy increases as we approach the boundaries of the universe.

Energy and Big Bang

Donal Goldsmith writes: "Particle physics proposes that during inflationary era, the universe acquired an enormous cosmological constant, which faded to zero as the universe became 10^{-30} second old."[29]

This is the De sitter model which claims that a tremendous amount of energy set off the initial inflation of the universe. Therefore, in De sitter model we just have energy, prior to the appearance of other elements in space-time. The familiar particles start to appear after that moment.

Energy and Microcosm

In small scale, we can write the Heisenberg Uncertainty relation again as:

$$\Delta E * \Delta T = h(\text{Planck Constant})/2\pi$$

$$\Delta E = h/2\pi\Delta T$$

As the changes of time gets smaller changes in energy and therefore the energy itself increases. When ΔT approaches 0 then ΔE approaches infinity. [40]

So at smaller time intervals, as we get closer to boundaries of space-time and Planck Sea the energy can increase up to infinity. As mentioned before zero point energy is believed to be a huge source of energy, which is extractable.

A) Energy and Quantum Mechanics
$c^2 = E/m$

In the Einstein' equation as mass decreases, energy increases. The amounts of energy for molecules are 10 to 10^{-3}ev. The energy of electrons bonds in atoms is 10ev. Inside the nuclei the energy goes up to 10^8ev. When particles are at rest the energy level is up to 10^{11}ev. And at the exit (Planck scale) the energy increases to an enormous amount of 10^{28}ev.[13]

B) Energy and Space-Time Singularities
When we pass the event horizon of black holes, as space and time shrink again, gravity crushes all in-falling masses. Later on

supposedly, matter will be decayed and will liberate energy, which will dissipate as radiation back to space-time. By definition, this is the way black holes disappear.

The above concept is in line with the wave-particle model that will be described in the next chapter. In my model, matter converts to energy upon exiting space-time and inside the black holes.

What if black holes are the big gates for energy exchange? Small gates being, Planck distances which are much smaller but infinitely more frequent holes.

Information

Information also has to increase as we get close to boundaries. The second law of thermo-dynamics indicates that in a closed system entropy increases while organization decreases with the passage of time. In an organized system, information is limited to the structural data of the system. On the contrary, in a structureless and chaotic system, information is maximal.

Information and Big Bang

If we bring the above principle to the time of The Big Bang when structure in the early universe was minimal, the information potentials should have been be maximal. As the universe becomes more structured the information is reduced and gets limited to the data of the existing structure of the space-time.

Information and Microcosm

The information at the proximity of internal borders of space-time increases rapidly. This can be observed in the ultra small (quantum mechanical) scale. In Compton Scattering, as Feynman's diagrams indicates, the possibilities of events are endless. It means that there are infinite data available for infinite possible variation of events.

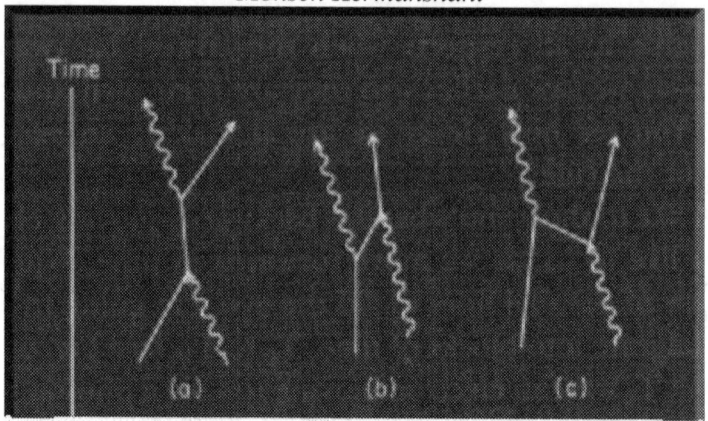

a few scenario of many posibilities of scatering after an electron
and a photon strike each other

To conclude it could be said that, as we move farther from the Planck arena and draw into space-time, the information diminishes. Conversely, in the smaller scales superposition of states induces maximum data. Therefore, at the inner boundaries of the universe we are faced with flourishing information.

Summary

In this chapter, I have demonstrated how gray zones appear at the inner and outer boundaries of the universe and how upon approaching these boundaries, the three elements of space-time go pale and information and energy gradually prevail. More evidences can be derived and presented to show how the intermingling of elements between the proposed singularity and space-time appears gradually around the boundaries.

138

Revisiting
Wave-Particle Function

Wave function was introduced as a model to describe the puzzle of The Heisenberg Uncertainty Principle. The uncertainty Principal claims that we cannot be certain of the location and momentum of a particle simultaneously. Nevertheless, we can adopt this model for the movement of free particles.

Complex Numbers

While looking at the quantum wave function from a mathematical point of view, Professor Roger Penrose indicates:

"You cannot explain the wave like nature of quantum particles in term of probability waves of alternatives. They are complex waves of alternatives!"[5]

In order to describe the wave particle function in this view, some knowledge about complex numbers is needed. Alternatively, just reviewing the assertions made in the complex number chapter should suffice to comprehend and follow topics that will be discussed.

As it was mentioned before, Complex Numbers are a combination of purely real and purely imaginary numbers.

Complex number = $(x + iy)$

Normally, we measure the elements of space-time by real numbers. The concept of the complex number implies that any of these elements should have an imaginary dimension in their nature.

Considering the notion of complex numbers, I have concluded that, the real axis value of elements in space-time periodically changes, disappears, and reappears. For example, if the real value in the X- axis denotes the mass of a particle, the mass has to appear and disappear in each period. This is what we see in the electrons around the nucleolus of each atom. The electron appears and disappears in a band like zone around the nucleolus. Erroneously, we call this zone orbit.

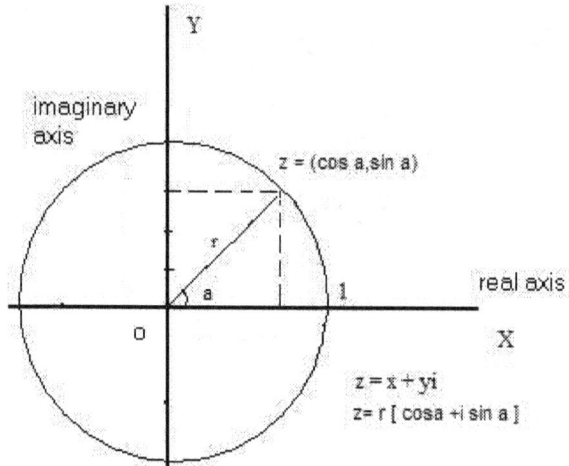

Resemblance of Argand Diagram and Trigonometry

In this model, we take the imaginary number (*i*) as a factor which represents the singularity effect on different phenomenon. So any mass as it intermingles (multiplies) with singularity alternates and shifts in between its real value and imaginary value. In other words, the particle intermittently disappears from space and reappears along its wave path. We can generalize this concept and conclude,

Assertion WP #1; while traveling through the fabric of space, objects take a journey and alternate between singularity and space-time.

In a complex number diagram, as the Z passes the first quadrant, it enters a new arena where the real value is negative (second quadrant). So, it does not completely disappear but it enters into another domain. How are we going to interpret this negative value?

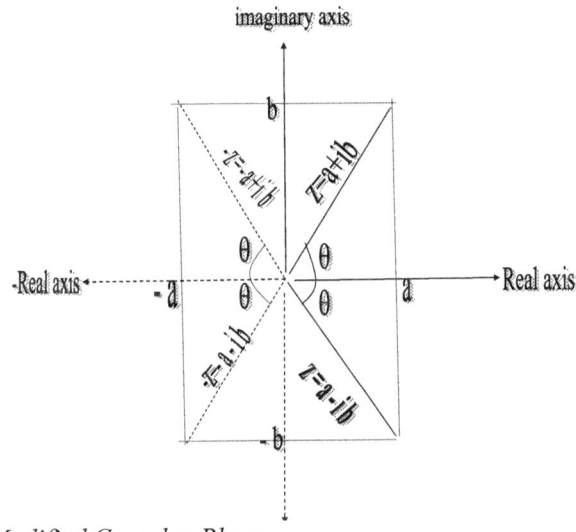

Modified Complex Plane

In the modified complex plane shown above, the real value reappears as the complex number rotates one $3/2\pi$ and enters the fourth quadrant. If x denotes the mass of a particle, we may interpret the negative domain in the complex number diagram where the particle loses its mass.

Compton Wavelength

Webster's dictionary provides conventional definition for wave as "A disturbance or variation that transfers energy progressively from point to point in a medium."

In 1924, Louis-Victor de Broglie postulated that *all* matterial objects have a wave-like motion. He postulated that any particle

Mohsen Kermanshahi

has an accompanying wave function. He also introduced Compton frequency, which is a kind of oscillation and circulation of the charge around a charged particle. Each object has its own and specific Compton Frequency.

He suggested the equation below to govern the relation between wavelength of a mass and its momentum *p*:

λ = h/p, where h is the Planck constant.

Any object has its own Compton wavelength.

Previously we speculated that the key to understand quantum mechanics is hidden somewhere in Campton wavelength of particles. *Wikipedia* describes the Compton wavelength as:

"The Compton wavelength λ of a particle *X* is given by $\lambda_X = h / m_X c$, where *h* is the Planck constant, m_X is the particle's mass and *c* is the speed of light. A particle generally behaves quantum mechanically when observed at distances shorter than its Compton wavelength."

Only Compton wavelengths of small objects such as atoms are detectable. The wavelengths of larger objects are so small that cannot be detected. Underneath we are going to analyze the particle's movement during its Compton wavelength. It furthers:

"In particular, in the uncertainty relation for position and momentum, $\Delta x \, \Delta p \geq h$, when the position uncertainty Δx is less than the Compton wavelength, the momentum uncertainty Δp is greater than $m_X c$. Since momentum carries energy, the uncertainty in energy is greater than $m_X c^2$, which is enough energy to create another particle of type *X*."

Therefore, the article predicts particle creation along the wavelength. This is in line with the scenario that we will be discussing in this chapter. The article further continues;

"The Compton wavelength is therefore generally viewed as the cutoff below which quantum field theory, which can describe particle creation and annihilation, becomes important."

In this model I will suggest creation and annihilation of particles along their Compton wavelength as well.

The Schrödinger equation for motion of particles (traveling wave) along *x*-axis at the time of *t* can be written as:

ψ (x,t)= A exp[*i (kx-ωt)*]

142

Where A is amplitude of the wave, k is wave number and ω is angular frequency. We may expand the above equation as:

ψ (x,t)= A exp[i ($kx\text{-}\omega t$)] =A cos ($kx\text{-}\omega t$)]+i Asin($kx\text{-}\omega t$)] .

This function is a complex number with i Asin$(kx\text{-}\omega t)$ as its imaginary part.

As we can see, on particle-wave function dimensions x *and* t are complexified (demonstrated in complex number version) by the imaginary number i. In order to describe the particle-wave function, physicists are not using the kaluza-klein dimension, or any of the extra dimensions in the string theory, instead they have to choose a virtual dimension out of our space-time to be able to explain the quantum wave function, an imaginary dimension. Yes, we can imagine the complex numbers in our mind. But wave function is happening all over the world in every moment, even if, our mind is not with it. Therefore, there should be another being out there to accept the image, to be able to accommodate the imaginary part of the complex numbers.

Wave Function in this Model
Mass, Energy-Information Phase Transition

Although wave function was introduced as a model to represent the Heisenberg Uncertainty Principle, we can make the following argument for a β ray (composed of electron particles). Einstein's Special Theory of Relativity says that nothing can travel at or above the speed of light in our space-time universe. The reason behind this is that in The Special Theory of Relativity, the relation between mass and its speed can be obtained from:

$m = m_0 / \sqrt{(1 - v^2/c^2)}$

Where m is the relativistic mass (mass in motion), m_0 is the mass of an object in rest, v is velocity of the particle relative to the observer, and c is the speed of light. Velocity increase, augments the particle's mass. At speed of light v^2/c^2 becomes 1. In this case 0 will divide m_0. So mass will increase to infinity.

$m = m_0 / \sqrt{(1 - v^2/c^2)}$

$$m = m_0 / \sqrt{(1-1)}$$
$$m = m_0 / 0 = \infty$$

Here, we return to the concept of infinities, the area that most physicists detest. This is another road sign indicating that we need to strive for a deeper understanding of reality. We must not ignore these signs that have been avoided for fear of the unknown). Since infinity does not exist in space-time universe, besides we need an infinite amount of energy to move this infinite mass, this goes beyond explained physics. Known physics does not have any explanation for it. John Earman writes:

"In 1960 Einstein found that, in a condition of static equilibrium, as the radius of the cluster approaches Shwarzchild radius. The particle would eventually have to move faster than light."[4]

By definition, there is no (mass-like) medium in space-time. The notion of "Ether" has been rejected at the beginning of the twentieth century. The mystery of fields (electromagnetic fields, Gravitational fields, etc.) has yet to be explained.

Let us suppose that during the course of its wavelength the particle actually oscillates and travels along a spatial dimension like X with linear speed v. We can take a bouncing ball as analogy. Please note that in such a wave motion the particle is traveling in another direction as well (along Y-axis).

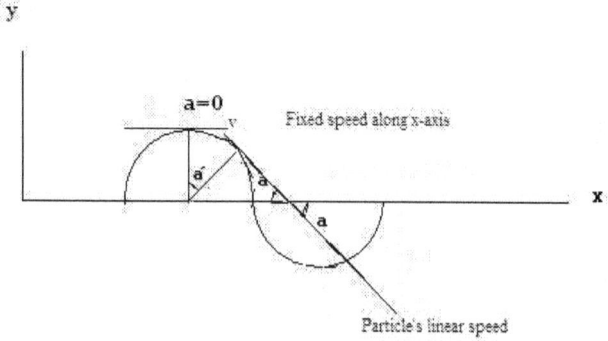

Acceleration of particle along the wave length

Diagram #1

The above diagram shows a polarized wave particle, which possesses mass and propagates along X-axis. If we take the propagation speed (v) as fixed, then the actual velocity has to change according to the position of the particle along the curve. The actual speed will exceed the propagation speed as the particle leaves the peak and travels down the slope. The equations can be written as:

$\Delta y / \Delta x = \tan. a$
$\Delta y = \Delta x \tan. a$
$\Delta y / \Delta t = \tan. a \, \Delta x / \Delta t$
$\lim_{\Delta t \to 0} \Delta t \, \Delta y / \Delta t = \lim_{\Delta t \to 0} \tan. a \, \Delta x / \Delta t$
If Δt gets near 0, $V y = \tan. a \, V x$

Please note that in the above diagram, the angles a and a' are equal because their sides are perpendicular to each other.
For the speed of particle in any point we can write:

$$V^2 = Vy^2 + Vx^2 = Vx^2 + Vx^2 \tan.^2 a = Vx^2 (1 + \tan.^2 a),$$

Therefore;
$$V = Vx \sqrt{(1 + \tan.^2 a)} \dots (1)^{40}$$

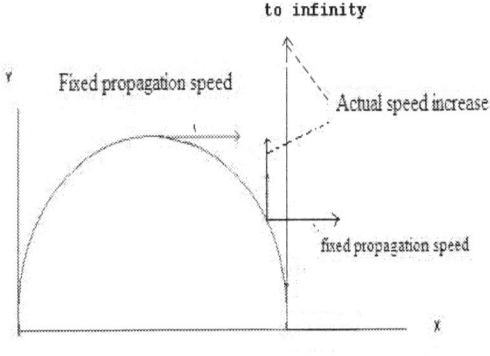

to infinity

Y   Fixed propagation speed

Actual speed increase

fixed propagation speed

X

Propagation of Beta wave in vaccum

Regardless of the propagation speed, as the particle approaches the x-axis there is a moment when its speed equals and exceeds the speed of light. Please note that for. $89°$ angle the tangent value s 57.2900 and its value for $90°$ is infinity. So for particle speeds, light speed is reached as it travels towards the X-axis. At $90°$ speed equals infinity.

$$V = Vx \sqrt{(1+tang.^2a)}$$
$$V = Vx \sqrt{(1+ \infty)}$$
$$V = \infty$$

Since infinity cannot be defined normally in space-time physics, we would say speed equals maximum. But because this model contains the singularity in which infinity is defined I chose to leave infinity in place.

Infinite Speed and Non-Local Zone

At the time when a = $90°$, Vx cannot have any amount other than zero in space-time. So for our original assumption to hold (Vx remains fixed all along the wavelength) we have to assume that the wave particle left leaves the space and enteres a zone with no dimension. With infinite speed (the assumed actual speed), the wave particle cannot remain in space-time. It has to leave and enter a non-local entity. We are led to this conclusion by the fact that the only place that an object can have infinite speed is a non-local zone.

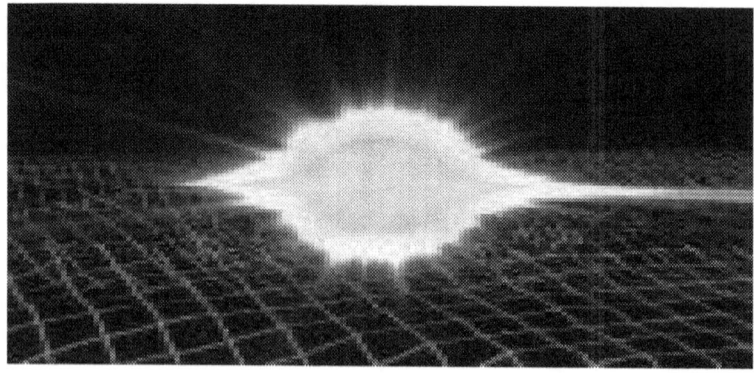

The space-time in this model is discrete, but singularity is in the neighborhood and readily available to deportees of space-time. But singularity in this model cannot contain mass. Therefore, considering the conservation of mass and energy law, we have to deduce that mass has been converted into energy while joining the singularity. According to the Standard Model of particles, subatomic particles are zero-size and mass-less. Higg's boson adds the property of mass to them while traveling in space-time (see Mass & Gravity chapter). Therefore, we may assume that the particles lose their mass while leaving space-time.

To build a more objective argument let us study particles with a propagation speed of 1/20c. If you are not up for the mathematics, you may rather skip this portion.

In equation #1, $V = V \times \sqrt{(1+\tan^2 a)}$. For the particle to reach to speed of light, the magnitude of a, can be calculated as:

$C = 1/20 \, c \, \sqrt{(1+\tan^2 a)}$, and,
$\sqrt{(1+\tan^2 a)} = c / c/20$, therefore,
$20 = \sqrt{(1+\tan^2 a)}$ then $400 = (1 + \tan^2 a)$

Thus $\tan^2 a = 399$ Hence $\tan a = 19.98$, which approximately corresponds to tangent of $87°$.

Invariance around X-Axis

We redraw diagram #1 for this speed as:

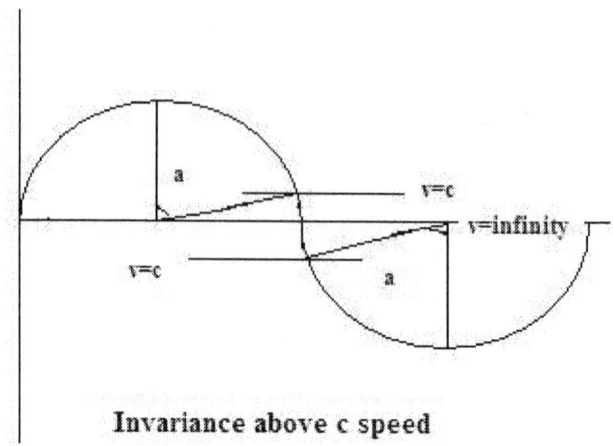

Invariance above c speed

Diagram # 2

Formula $V = V \times \sqrt{(1+\tan g^2 a)}$ requires that as the *a* increases beyond $87°$ the speed also increases and at
$a = 90°$ the speed of the particle-wave reaches to infinity. The calculations below show that beyond *c* speed, the elements under examination do not change.

We formerly assumed that at speed of light mass has to convert to energy. The amount of mass itself calculated from $m = c^2/E$. In addition, the momentum (p) of the wave-particle at this point is obtained from:

$P = (E/c^2)*V$

As the wave-particle departs from $87°$ and gets to $90°$ (hits x-axis), tang.a will be infinity so speed will amount to infinity. If V= infinity, then:

$P = (E/c^2)*V = (E/c^2)*V = (E/c^2)*$ infinity = infinity

The momentum equals infinity and:

E total $= \sqrt{(p^2 c^2 + m_0^2 c^4)}$ = infinity

It is apparent then that in our example from the 87^0 to the 90^0 zone the amount of energy remains infinite.

Mass at 90°

Let us look at mass at 90^0. For mass at 90^0 angle , we may write,

$m = m o / \sqrt{(1 - v^2 / c^2)} = m o / \sqrt{(-1 (v^2 / c^2 - 1))} = m o / i^2 (v^2 / c^2 - 1) = m o / i^2 (\text{infinity}^2 / c^2 - 1) = 0.$

This denies the presence of mass in the 90^0 zone.

Space around X-Axis

If momentum is infinite in 90^0 , then space looses its meaning around the X-axis:

$\lambda = h / p = h / \text{infinity} = 0$

And the wave number k turns to $k = 2\pi/\lambda = 2\pi/0 = \text{infinity}$.

$\Delta k = k - k_o, \quad \Delta k = \infty - k_o, \quad \Delta k = \infty$

In addition, we can write the Heisenberg uncertainty formula as:

$\Delta k * \Delta x \geq h/2\pi$

Therefore Δk has an inverse relation with Δx

$\Delta x \geq 1 / \Delta k$

If Δk equals infinity, then Δx can shrink to zero, which means, there is no movement in space. We can interpret and speculate that wave-particle does not encounter space at $a = 87^0$ to 90^0.

As an alternate approach, at the infinite energy point we might express the following:

$\psi (E) = (E \max) e^{i (kr^n - wt)} = (E \max) e^{i (kr^n - wt)}$

E = infinity = (E max), then, $e^{i(kr^n - wt)} = 1$ Therefore $i(kr^n - wt) = 0$

Or, $k\,r^n = wt$, And since $2\pi/\lambda * r^n = 2\pi/T * t$, Hence, $r^n = t * \lambda/T = 0$. We conclude that there is no space in $90°$ point. Then we can conclude that,

Assertion WP #2: space disappears intermittently during the course of the particle's journey along its wavelength.

Time around X-axis
We can also write the Heisenberg Uncertainty Principle as:

$\Delta t * \Delta E \geq h/2\pi$
If ΔE = infinity then, $\Delta t = 0$

Again, it shows that time remains put for wave-particle at $a=87°$ or $90°$.[40]

Assertion WP#3: Time stops intermittently for the objects during the course of their wave motion.

Second Half of the Wavelength
Then, the particle leaves the x-axis and moves towards the trough. Here, the changes in the Y-component of the movement decelerates as does the linear speed until the particle surpasses $87°$ again and reaches a speed that it can have in space-time (less than c). This is the time for the particle to acquire mass again. The particle reappears when the wave reaches the possible speed in space-time. This can be an explanation for the experiments which show electron jumping in and out of the existence in our universe.

Bouncing Ball
The assumed function mimics the bouncing ball in a frictionless gravitational field. Let's look at bouncing ball scenario.
Gravitational force pulls the ball to the ground. When the ball hits the ground, the electromagnetic force of the atoms in the outer layer deflect the ball. As a result the ball bounces back.

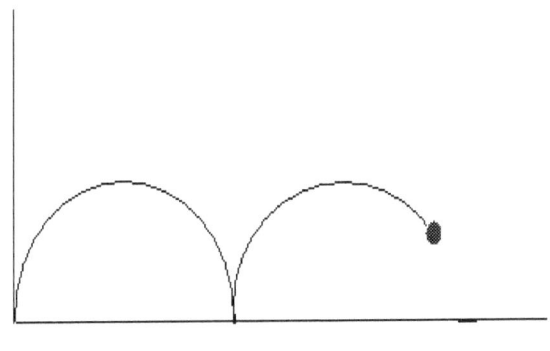

Bouncing Ball

Friction-less gravitational field

For an object to have a wave like motion, the interface between proposed singularity and space-time must possess an attractive and deflecting force. The uncertainty principle predicts that in small scales the gravitational fields rise and fall. Mind you that, according to the General Relativity, the shape of space follows the changes in the gravity. If the space in ultra short scale is curved the object will follow the curved space in a wave like motion.

On the other hand, the energy in singularity has to be unified and invariant. (Note 1)

Heisenberg Uncertainty Principle

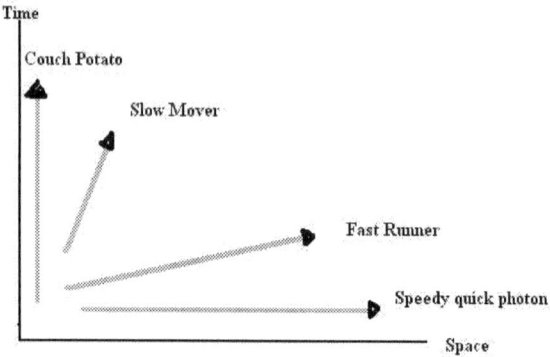

Combined movement through space and time dimensions

According to the Special Relativity theory, a stationary person just moves through time dimension. Thus gets older fast. A person moving with speed of light only travels in space dimension. Therefore stays young forever. There is no passage of time for such a person. Other movers moving through a combination of space and time according to their speed.

In addition, the combined speed of an object through space and its motion through time is precisely equal to the speed of light.

Going back to particle's wave function in this model, we can sketch the diagram below to illustrate particle's motion through the space and time.

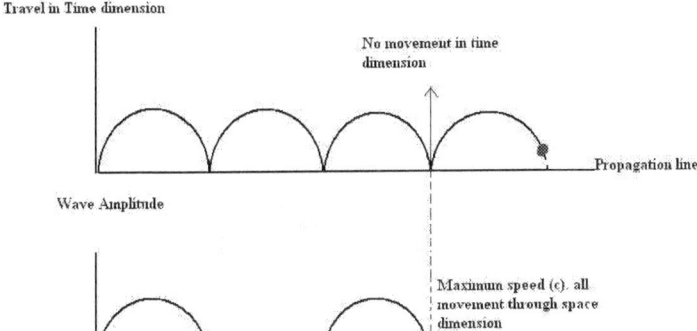

Closer to the X-axis, the particle's speed equals light speed. Therefore, its movement through time dimension is ceased. According to the Heisenberg Uncertainty Principle, the relation between energy (E) of a particle and time (t) is obtained by:

$$\Delta E \; \Delta t \geq h / 2\pi$$

$$\Delta E \geq h / 2\pi / \Delta t \geq h / 2\pi \Delta t$$

Where h is Planck Constant. When the particle is around the x-axis, we pin point time at zero. At this point, time changes are also amount to zero. Heisenberg equation indicates that when the changes in time is zero the changes in energy of the particle increases to infinity. This implies that the actual energy magnitude can amplify up to infinity as well. Therefore, we can conclude that around x-axis the particle enjoys the infinite energy.

On the other hand, for the Uncertainty between momentum (p) and position (x) we can write Heisenberg equation as:

$$\Delta p \; \Delta x \geq h / 2\pi$$

The diagram below shows the location and momentum relation in our model.

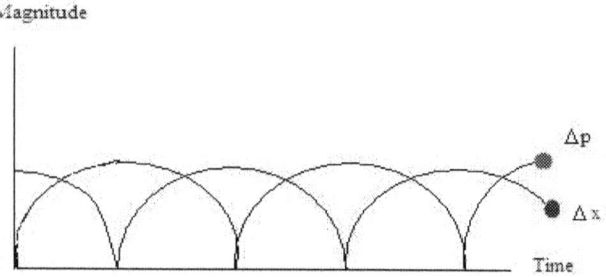

Heisenberg Uncertainty Principle
Momentum versus Location
Bouncing ball model

In our model, when a particle moves down the slope from the peak during its wave function, the changes in position along x-axis (Δx) are diminished. However, the changes of its speed and therefore its momentum increase.

At the peak, the particle has the least momentum and changes of momentum. Around the x-axis, the particle's displacement is zero, therefore the changes in the momentum is maximum. As the particle travels down the slope, its momentum increases and the rate of increase multiplies as it closes to x-axis. More momentum means more energy. So closer to x-axis, the time changes are minimal and energy changes are maximal. This property of our model is in line with the Heisenberg Uncertainty principle for time and energy.

Max Born Prediction

In 1926 Max Born suggested that:

"An electron wave must be interpreted from stand point of probability. Places where the magnitude of the wave is large are places where the electron most likely to be found. Places where the magnitude is small, are places where the electron less likely to be found."[1]

The Born's electron probability wave favors the presence of the electron near the peak of the translation wave. Here we can offer an explanation for the Max Born's suggestion. In our wave model, the speed of electron s reduced around the peak of translation wave, therefore it is detectable in space-time. However, we can assume that as the actual velocity increases to reach the light speed (299,792,458 m/s) the probability to pin point it is reduced. At the speed of light electron as a mass disappears from space-time. Therefore, we cannot follow its trajectory.

Wave Function and Complex Numbers System
According to Lorentz equation the length of a particle decreases at higher speeds (length contraction). We may write Lorentz transformation equation for linear motion as:

$$L = L_0 \sqrt{(1 - v^2/c^2)}$$

In velocities greater than the speed of light (v>c) Lorentz transformation equation moves us to the imaginary domain because $(1-v^2/c^2)$ turns into a negative number. $L = \sqrt{-n}$. [40]
Formerly the imaginary domain was related to the proposed singularity; hence I conclude that, any particle which exceeds the speed of light moves to singularity.
The above interpretation of wave function offers an explanation for using complex numbers to define wave function in the Schrodinger's equation mentioned above. While real number corresponds to the particle in space-time, imaginary number relates to the trace of particle in singularity. That is one way to describe wave function by complex numbers.

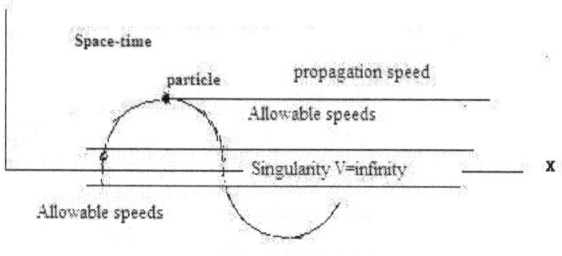

Modified C Diagram

The above diagram shows were the objects meet the singularity during their wave function.

Water Waves

Water waves can be taken as an analogy for the above concept. As the water molecules rise from the surface we see them as wave. When they fall, they join the sea of water again. And then the cycle repeats itself. What we see are waves at the surface, but in reality, the waves are extension of the sea. They take a specific shape while in action. Please note that the identity of the molecules of water in subsequent wave is not necessarily similar to previous ones. This resembles the identity problem of particles, in quantum fluctuation. By definition, we cannot identify different electrons from each other. We just see the wave.

The main factors, which are hidden here, are the energy field, which is moving the water molecules and the data, which regulates the movement and shapes of wave's motions. The collective action of field (data and energy), which is not observable, and the water molecules, which are observable, form the visible waves. Just getting preoccupied with the shape of the waves and ignoring and normalize contributory factors from underlying sea is not a sound strategy.

We can mention the musculo-skeletal motion as explained in Holonomic Brain Theory as an analogy, as well. In this theory, the movement footprint is in spectral and non-local form. However, when the behavior is materialized it turns to body motions, which are local and observable. Mind you that mind is one of the main contributory factors in this model.

Dirac's Electron
The Dirac equation for electron can be written as:

$$\Psi = (\alpha\, A\, ,\, \beta\, A'\,)$$

This represents a pair of 2-spinors. We can interpret its physical reality as follows. An electron actually consists of two separate particles (α A and βA'). These two particles have opposite charges and are continually converting into one another.

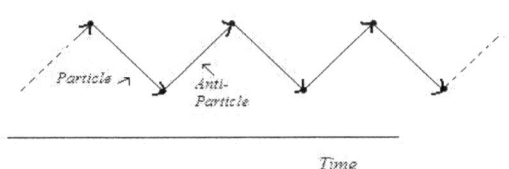

Dirac Electron
Particles constantly convert to each other

Roger Penrose, Road to Reality[56]

In Dirac's electron wave model, the anti-particle in second portion of the phase, has an opposite charge and is called anti-particle. The anti-particle of electron is positron. The birth and rebirth of the Dirac's electron is in line with this model. However, the instantaneous speed of Dirac particles is always constant and equal to the speed of light. The variation in propagation speed comes from the zigzag motion of two components as they average out. In this model, I have proposed that the speed along the propagation line is constant but the instantaneous speed of the particle changes between the propagation speed and the speed of light in a 2-dimensional wave plane. The disappearance and rebirth of the particle comes by its entering and rising from the singularity.

Super-Symmetry Theory
The above definition for the wave-particle function, which suggests a correlation between wave-particles and singularity during the course of action, resembles super-space in the super-symmetry theory. In Super Symmetry's super-space, bosons and fermions (two main categories of sub-atomic particles) can interchange without affecting the theory. Interestingly, extra super-space dimensions in super-symmetry do not have any size which mimics the proposed singularity. Also, in the proposed singularity, bosons and fermions are treated symmetrically.

Zero Point Energy
From here the stochastic electrodynamics will continues to explain the scenario. The theory is based on assuming that zero point energy (ZPE) is pervasive throughout the space. This energy is supposedly an integral part of the universe. The theory can explains many physical paradoxes just by adding the concept of ZPE to the ordinary classical physics. For example, it can explain numerous phenomena including inertia, gravity and the radiation paradox of the Bohr's atom based on fluctuating electromagnetic field associated with zero-point energy. Previously, we attributed zero point energy to singularity. According to stochastic Electrodynamics, the ZPE field creates a range of superimposed random waves of all frequencies and phases in all directions, with a power spectrum proportional to the cube of frequency.

Haisch, Rueda and Puthoff (HRP, 1994) in the paper "Inertia as a zero-point field Lorentz force" showed how inertia (the force needed to alter an objects movement) can be explained by interaction of particle with zero point fields. They assumed that:

"A fundamental particle (such as an electron) could be treated as a two-dimensional Planck oscillator driven by electric components (Ezp) of the ZPF to oscillate in the xy-plane. They then examined the effects of the magnetic components (Bzp) of the ZPF on the Planck oscillator under the condition of constant acceleration in the z-direction. The result was that the Lorentz force due to Bzp fluctuations proved to be proportional to the acceleration of the Planck oscillator, thus suggesting its interpretation as the reaction force due to inertia."[54]

Combining the above explanation with the presented model for wave-particle, we find a good explanation for equivalence principle (Which implies that mass and gravity are equivalent with each other). I refer the reader to California Institute for Physics and Astrophysics web site for further information about Stochastic Electrodynamics.

Tachyon

This model also incorporates and explains the notion of tachyons. A tachyon is a hypothetical particle that travels at superluminal velocity. By definition, tachyon's mass squared is negative. That means to define it we have to use imaginary number ($\sqrt{-n}$). Alternatively, we may say its rest mass is imaginary. We may also re-word the previous sentence by saying; tachyons rest mass belongs to informational domain. Previously, we considered the singularity as information domain. We also speculated that faster than light speed objects loose their real mass and enter the singularity.

Nature of Photons

Same speculation can be extended to photon itself. By definition, photon does not posses any mass. The speed of photon as light particle is constant and amounts to approximately 300.00km/s. As such, it has to reside in the boundaries of space-time. As mentioned in boundaries chapter at the vicinity of borderline matter gets pale and disappears. Therefore, since photon stays in the boundaries it does not possess mass.

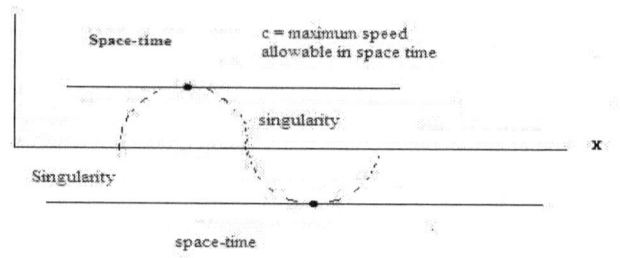

Photon projection along X-axis
No acceleration in space-time

Photon stays at the boundaries of space-time

Maybe that is why Pauli Exclusion Principle which indicates that two similar particles cannot occupy the same location in space, does not apply to photons. Since Photon stays at the boundaries but does not enter space-time, there is no competition for obtaining territory. There is no territory at all to fight for.

M Theory

The M Theory (the new and united version of string theories) explains the process of the appearing and disappearing particles by postulating that particles are absorbed and emitted by the p-brane. The p-brane is a hypothetical space-time, which can have different dimensions. The advantage of the p-brain assumption is that it keeps us in our comfort zone while providing solutions for our paradoxes. In the model presented in this chapter, the transformation is happening in an entity outside of the space-time.

Defiance of Heisenberg Uncertainty Principle

My description of particle wave function is also in accord with recent experiments conducted by Shariar S. Afshar from Rowan University in New Jersey.

Afshar defies the notion that nature never permits us to observe both the particle and wave aspects at the same time. The uncertainty principle conveys that a particle can be either a wave or a particle and never simultaneously both. In the above model, I have described the particle as a real entity which is moving in a wave like motion, capable of being detected in both versions simultaneously, at some occasions.

Singularity: An Active Entity

In order to uncover solutions for the paradoxes, singularity cannot be considered merely an inactive pool of energy and information. The particle wave is continuous.

Below zero is not nothingness. There exists a vast computational territory, territory of negative numbers. Interestingly, these numbers interact with positive numbers and affect them drastically. Contextuality in the Bell-Kochen-Specker theorem can be viewed as evidence for such activities. The theorem simply states that in quantum mechanics, the value of different non-compatible observables is also correlated with each other, even though they are not complimentary pairs (complimentarity characteristics like location/momentum and energy/time. See Quantum Mechanic chapter). For the values to be interrelated we need an active media which not only accommodates the values, but also intermingles them.

In wave motion, the particle reappears and carries the information back to space-time. Quantum tunnelling (when a particle passes through a barrier and only reappears in space-time when the barrier is out of the way) can be explained by assuming the presence of a particle in singularity. Remember that particles in the Standard Model (standard model of particles is a science which deals with subatomic particles) are zero-size.

If we have in depth and precise knowledge about the true nature of things, we will come to the conclusion that "God does not play dice with the universe". We will come back to a deterministic world again.

The above model can be applied to wave-particles, such as alpha and beta rays or atoms and objects that are even more massive. According to Louis de Broglie, any object has a wave. However, massive objects have much smaller wavelengths. Cohesion of a system requires that each component resonate at the same phase. In other words, the whole assembly should have the same wave function. Quantum biologists believe that organisms have macroscopic wave function. The wave demonstrated can be a harmonic of a much smaller microscopic wave that according to de Broglie every object including live organism owns.

Therefore, it can be speculated that we as humans are also have a wave. Can we further assume that we are also traveling to singularity back and forth many times per second? The 16th century philosopher Mulla Sadra proposed a similar idea under the name of *al-haraka al-jawhariyya*.

Substantial Motion

Mulla Sadra (1571/2-1640) The Iranian philosopher, and perhaps the single most important and influential philosopher in the Muslim world maintained in his substantial motion (*al-haraka al-jawhariyya*) that:

> "Substance only changes suddenly, from one substance to another or from one instant to another, in generation and corruption."[62]

This is in line with the complex number assertion C#1, which postulates that matter appears and disappears periodically.

It is also in line with Assertion WP#2 in this chapter, where it is suggested that mass joins singularity and appears again in space-time in each Compton wavelength.

Notes

1) Here, I am referring to Poynting vector of the zero-point fluctuations. Interested readers can look at the following website for further information:

http://en.wikipedia.org/wiki/Poynting_vector

Mass and Gravity

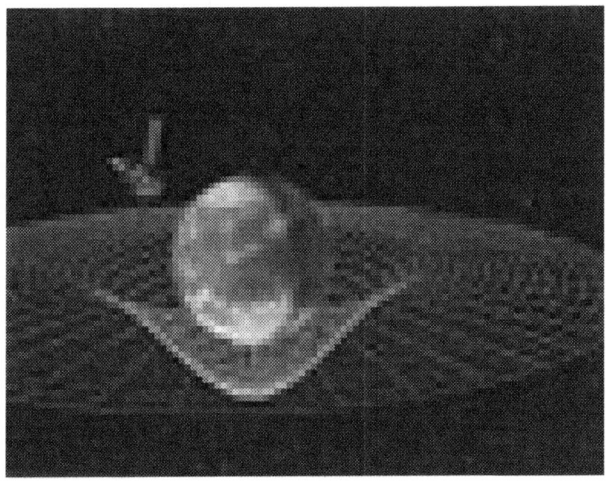

http://www.nasa.gov/

Gravity is the central dilemma of the theory of everything, and it so happens that it is the central chapter of this model as well. In addition, I will offer an explanation for the unexplained Planck Constant in the paragraphs to come. An explanation is also proposed for the nature of the mass of subatomic particles. The explanations are verified by experimental findings in nuclear physics and the standard model of sub-atomic particles.

Mass

By definition mass is the resistance of a particle to acceleration. This property is also called inertia. The true nature of mass is not completely understood.

The standard model of particles postulates that sub-atomic particles by themselves do not posses mass. Mass is defined only by inertia. The Newton equation for inertia is $F = ma$, where m is mass of an object, F is the force needed to accelerate it with the magnitude a.

Gravity

Newton's law of universal gravitation denotes that all objects attract each other. The force of attraction is directly related to the object's gravitational mass.

It is astonishing that these seemingly different properties of objects, namely inertial mass and gravitational mass, are equal. This phenomenon is called the Equivalence Principle.

At this point, let us look at the main existing theories, which try to explain inertia and mass.

Higgs Mechanism

The most popular belief for a description of the origin for mass is the Higgs Mechanism. As mentioned before, the standard model of particles postulates that particles by themselves do not posses mass. The equation for mass acquisition through Higgs mechanism is given by:

$$m_i = \Gamma \, \hbar\omega^2_c/2\pi c^2$$

Where Γ is Abraham –Lorentz damping constant. $\hbar$ is Dirac's constant and ω_c is cutoff frequency (the frequency at which a mass respond to and starts oscillating). Please note that, everything else being fixed, in Higgs mechanism, the mass is directly related to the cutoff frequency of the wave-particle.

The Higgs postulate assumes a universal field called Higgs field that is carried by the Higgs Boson. Higgs Boson is a hypothetical particle that supposedly introduces mass to other particles through Higgs mechanism.

"The Higgs idea comes directly from the Physics of Solids. A solid contains a lattice of positively charged crystal atoms. When an electron moves through the lattice, the atoms are attracted to it, (therefore slowing it down) causing the electron's effective mass to be as much as 40 times bigger than the mass of a free electron."[64]

This concept has been extended to define the nature of mass acquisition by particles.

The Higgs particles supposedly create crowding and traffic in a particle's way. Although they are not affecting the homogenous motion of particles they some how resist against particles acceleration. This is proposed as the reason behind resistance of the particle to change of trajectory and acceleration. So far, hard work in the Cern and Fermi lab and other accelerators and colliders throughout the world failed to find Higg's boson. In fact, the Dec 2001 issue of the New Scientist published an article titled "No sign of the Higgs boson" with a strong suggestion that the Higgs boson does not exist.

Even if we find such a boson, we still have to find an explanation for the way that it actually creates inertia.

David Miller from Dept. Physics and Astronomy, University College London says: "A crystal lattice can carry waves of clustering without needing an electron to move and attract the atoms. They are called phonons"

Dr. Miller continues, "there could be a Higgs mechanism, and a Higgs field throughout our Universe, without there being a Higgs boson."[64]

The amount of energy added by Higgs field is:

$$E = M^2h^2 + Ah^4$$

Where A is a positive but unknown constant, h is the size of Higgs field, and M is the mass of Higgs particle.

The researchers at California Institute for Physics and Astrophysics question Higg's postulate of mass acquisition as:

"Why does the energy "soaked up" from the Higgs field resist acceleration? Perhaps that is not a legitimate question. Perhaps mass and energy intrinsically possess the property of inertia and that is the end of the story."[65]

Dirac's Electron and Higgs Mechanism
The Dirac equation for electron is obtained by:

$$\Psi = (a_A, b_{A'})$$

It represents a pair of 2-spinors. We can interpret its physical reality as follows. An electron actually consists of two separate particles (a_A and $b_{A'}$). By definition $b_{A'}$ is an antiparticle. These two particles have opposite charges and continually convert to each other. In case of electron, the anti particle is called positron. Positron has been actually detected.

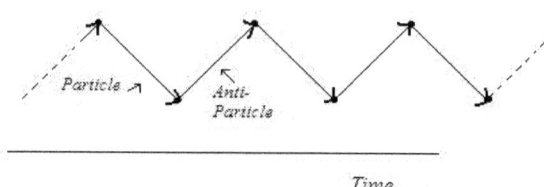

Dirac Electron
Particles constantly convert to each other

Roger Penrose, Road to Reality[56]

The conversion coupling constant is $2^{-1/2}$ M. The quantity M = $\hbar$/ μ where $\hbar$ is Dirac's constant and μ is the rest mass. In Higgs view $2^{-1/2}$ M is a field where particles drop and acquire mass again.

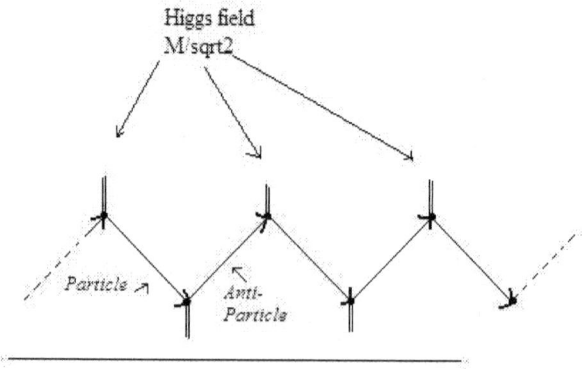

Dirac Electron and Higgs Field
Particles constantly convert to each other while droping and acquiring mass from Higgs Field

Stochastic Electrodynamics

Semi-classic stochastic electrodynamics has been well developed since 1960. It postulates that at a microscopic level, the field is filled with numerous plane waves that extend in every direction. It hypothesizes that the waves come from the Zero Point Field (ZPF).

"Stochastic electrodynamics postulates that the ZPF is as real as any other radiation field. In such a view the existence of a real ZPF is as fundamental as the existence of the universe itself. The only difference between stochastic electrodynamics and ordinary classical physics is the single assumption of the presence of this all-pervasive, real ZPF, which happens to be an intrinsic part of the universe."[65]

Knowledge about zero point energy is needed to comprehend the mass explanation in The Stochastic electrodynamics view. Thus, I will elaborate more about ZPE in following paragraphs.

Zero Point Energy

As mentioned before, quantum mechanics predicts the existence of Zero Point Energy. The Casimir effect and Lamb shift phenomena are evidence for ZPE. Additionally, the one-dimensional harmonic oscillator also argues in favor of its presence.

"The theory of electromagnetic radiation is quantized by treating each mode as an equivalent harmonic oscillator. From this analogy, every mode of the field must have hf/2 as its average minimum energy."[65]

The phenomenon has been described in Singularity chapter.

P.S. Wesson from the University of Waterloo among other physicists infers: "Search into zero-point physics is justified and should be supported"[66]

At least on paper the use of ZPF to extract energy is possible. In fact Haisch and Rueda have presented a paper to NASA conference on space craft propulsion in 1997, suggesting the use of ZPF for future space drives.[67]

The followings are the beliefs of the researchers in California Institute for Physics and Astrophysics about ZPF and stochastic electrodynamics:

"In fact, two distinct views about it exist today.

One justification for making such an assumption is that by adding the ZPF to classical physics many quantum phenomena can be derived without invoking the usual laws or logic of quantum mechanics. It is premature to claim that all quantum phenomena could be explained by stochastic electrodynamics (that is, classical physics plus the ZPF), but that claim may one day turn out to be the case. In that event, one would have to make a choice. One could accept the laws of classical physics as only partly true, with a wholly different set of quantum laws required to complete the laws of physics; that is essentially what is done in physics now. Or one could accept the laws of classical physics as the only necessary laws, provided they are supplemented by the presence of the ZPF"[66]

According to Equivalence Principle, if the ZPF gives rise to the phenomenon of inertia, it must also generate the effect of gravity in some way.

There are outstanding issues regarding ZPE; for instance, if there is such a field in space-time, the gravitational effect must be enormous.

If we take ZPE as an internal element to space-time, its gravitational effect would be enormous. Paul Wesson writes:

"It is also claimed that if the ZPF really exists, it would be such an enormous source of gravitational force that the radius of curvature of the universe would be several orders of magnitude smaller than the nucleus of an atom. Of course, such a conclusion directly conflicts with everyday experience. The fallacy in the argument is that in the Sakharov-Puthoff model the ZPF as a whole would not itself gravitate. The gravitational force results from perturbations of the ZPF in the presence of matter. In the Sakharov-Puthoff model, then, the uniform ZPF is not a gravitational source and hence would not contribute to curving the universe."[66]

For more detailed information about ZPF please check http://www.calphysics.org/.

In addition, this field must affect the electromagnetic radiation wavebands. This effects have not been observed.

Quantum Vacuum Inertia Hypothesis

Quantum Vacuum Inertia Hypothesis mainly developed by California Institute for Physics and Astrophysics (CIPA). CIPA takes (ZPF) as virtual because if real it has to have certain cosmological effect that is not consistent with observations. The hypothesis speculates that the electromagnetic quantum vacuum (ZPF) contributes to the inertial mass of a matter. The nature of this effect is explained under the name Rindler Flux (Please check http://www.calphysics.org/rindler.html). In Rindler frame model the resulting force is proportionate to acceleration.

Rest Mass in Quantum Vacuum Inertia Hypothesis (QVIH)

Since a particle continuously interacts with the ZPE fluctuations, it exhibits Brownian-like motion. In the CIPA interpretation, this is the origin for quantum foam. A tiny bit of the quantum vacuum energy is diverted into the kinetic energy. This gives the Compton frequency and therefore the rest mass to the particle. In the CIPA view, "one could think of a particle as a localized concentration of zero-point energy which gravitates and resists acceleration"[65]

Inertia in QVIH

From Newton's second law of motion we know that a stationary mass (m) subject to a force (f), will accelerated (a) in the direction of the force. This acceleration is proportional to exerted force. F =ma.

The inertial mass in CIPA model is described as the effect of ZPE on accelerating objects. The accelerating objects interact with electromagnetic random waves of ZPE described above. The movement generates a drag force that is proportionate to acceleration. In CIPA view, this drag force is the origin of inertia and it is called acceleration-dependent drag force.

Therefore, In CIPA model, when force is applied to an object, it prevents it from following its own trajectory. This is called inertia and is the origin of the notion of mass.

Particle's Frequency

In (CIPA) model, the particle is actually traveling along the wave that is a geodetic trajectory in space-time. The frequency of particles depends on the energy of wave that carries it. So a force is needed to prevent the object from following its trajectory.

"The waves are fact ripples in space-time. These waves do carry energy, and each wave has a specific direction, frequency and polarization state. This is called a "propagating mode of the electromagnetic field."[65]

In this model the waves are present and a particle by getting involved in any specific wave obtain the specific energy and frequency. The frequency identifies the nature of the particle. This is the way that particle get their identification. Authors Bernard Haisch, Alfonso Rueda, L. J. Nickisch, Jules Mollere postulate:

"Zero-point fluctuations give rise to space-time micro-curvature effects yielding a complementary perspective on the origin of inertia. Numerical simulations of this effect demonstrate the manner in which a mass-less fundamental particle, e.g. an electron, acquires inertial properties."[70]

One may ask why there are just three main stable fermions (up quark, down quark, and electron) and a handful of unstable particles. Why the mass of proton and neutron are not proportionate to their constituents whereas, bigger objects mass is incremental? CIPA researchers believe,

"The quantum vacuum inertia hypothesis strongly suggests that the interaction between the quantum vacuum and charged fundamental particles (quarks and electrons) takes place at specific frequencies or resonance."[65]

Gravity in Quantum Vacuum Inertia Hypothesis (QVIH)

We feel and observe the effect of gravity constantly. However, the nature and dynamics of gravity is not well understood. The mainstream physics classifies gravity as one of the four main forces of nature (along with electromagnetic, strong and weak nuclear forces). Uphill efforts to unify the gravity with three other forces (grand unify theory) has not been fruitful so far.

Dr. H. E. Puthoff one of the researchers of CIPA describes the nature of gravity from the QVIH standpoint:

"Taking a completely different track ... the well-known Russian physicist Andrei Sakharov put forward the somewhat radical hypothesis that gravitation might not be a fundamental interaction at all, but rather a secondary or residual effect associated with other (non- gravitational) fields.

Specifically, Sakharov suggested that gravity might be an induced effect brought about by changes in the zero-point energy of the vacuum, due to the presence of matter. If correct, gravity would then be understood as a variation on the Casimir theme, in which background zero-point-energy pressures were again responsible. Although Sakharov did not develop the concept much further, he did outline certain criteria such a theory would have to meet such as predicting the value of the gravitational constant G in terms of zero-point-energy parameters."[22]

Dr. Puthoff and CIPA researchers followed the above lead and developed it further. Their efforts provided the following positive results.

"The gravitational interaction is shown to begin with the fact that a particle situated in the sea of electromagnetic zero-point fluctuations develops a "jitter" motion, or ZITTERBEWEGUNG as it is called. When there are two or more particles, they are each influenced not only by the fluctuating background field, but also by the fields generated by the other particles, all similarly undergoing ZITTERBEWEGUNG motion, and the inter-particle coupling due to these fields' results in the attractive gravitational force.

Gravity can thus be understood as a kind of long-range Casimir force. Because of its electromagnetic underpinning, gravitational theory in this form constitutes what is known in the literature as an "already-unified" theory. The major benefit of the new approach is that it provides a basis for understanding various characteristics of the gravitational interaction hitherto unexplained. These include the relative weakness of the gravitational force under ordinary circumstances (shown to be due to the fact that the coupling constant G depends inversely on the large value of the high-frequency cutoff of the zero-point-fluctuation spectrum); the existence of positive but not negative mass (traceable to a positive-only kinetic-energy basis for the mass parameter); and the fact that gravity cannot be shielded (a consequence of the fact that quantum zero-point-fluctuation "noise" in general cannot be shielded, a factor which in other contexts sets a lower limit on the detectability of electromagnetic signals)."[22]

The CIPA conjecture for the nature of gravity force between two objects is:

"The secondary electromagnetic fields turn out to have a remarkable property. Between any two particles they give rise to an attractive force. The force is much weaker than the ordinary attractive or repulsive forces between two stationary electric charges, and it is always attractive, whether the charges are positive or negative. The result is that the secondary fields give rise to an attractive force we propose may be identified with gravity."[65]

The CIPA researchers believe that inertial and gravitational mass are the identical thing. Inertia is felt as an object accelerates through the electromagnetic quantum vacuum. The gravitation is actually acceleration of the electromagnetic quantum vacuum past a fixed object.

"The latter case occurs when an object is held fixed in a gravitational field and the quantum vacuum radiation associated with the freely-falling frame instantaneously co-moving with the object follows curved geodesics as prescribed by general relativity."[70]

For detailed and updated description of Quantum Vacuum Inertia Hypothesis Please check reference #65.

ZPE in our Model

There are similarities between what I am going to present here and the CIPA model, because in both models ZPE is the origin for mass and gravity. In our model though, the origin of the ZPE is esoteric (external) but the resulting fields are intrinsic of the universe. So there will be some fundamental differences.

If we take ZPE as being out of space time and accept it as a property of proposed singularity. Then it can provide answers to questions like why the it does not affect the wave bands such as isotropy of microwave back ground, infra-red, optical and ultraviolet rays.

The particles carry a minute portion of the external ZPE inside the space-time.

Rest Mass in this Model

Einstein's Special Relativity implies that a particle's mass in motion increases proportionate to its speed. Lorentz equation for transforming mass in motion is:

$$m = m_0 / \sqrt{(1 - v^2 / c^{2})}$$

Where m is mass in motion, m_0 is the rest mass. V is the speed of object and c is the speed of light. Because the speed is the main variable here, is it fair to conclude that the nature of mass has something to do with kinetic energy of the object?

173

Underneath I quote from Professor Wesson of university of Waterloo and take the lead from following statement:

"Haisch and Rueda (1999a) returned to the issue of non-linearities, arguing that the observed masses of particles (e.g., the electron mass at 512 keV) are due to resonances in the electromagnetic ZPF. They also suggested that the scattering of the ZPF by a charged particle takes place at the Compton wavelength ... and that this leads to the de Broglie relation characterizing the wave description of the particle in terms of $deB = h = p$ (where p is the momentum and h is Planck constant.). This extension of their previous work is interesting; but in terms of making contact with the testable aspects of wave-mechanics, needs to be extended to a full discussion of the wave"[66]

In 1920, De Broglie proposed that every object has a wave like motion. In accordance with De Broglie's postulate, larger objects follow shorter wavelength and have higher frequencies. Quantum mechanics on the other hand tells us that the smaller the scale the more turbulence is observed in fabric of space.

In previous chapters, we postulated that in sine wave motion a real particle actually moves along a wave. In this scenario, if we take the speed along the x-axis, fixed at all times the actual velocity of the particle increases with the raise of the tangent of the angle a.

$$V = Vx \ tang. \ a$$

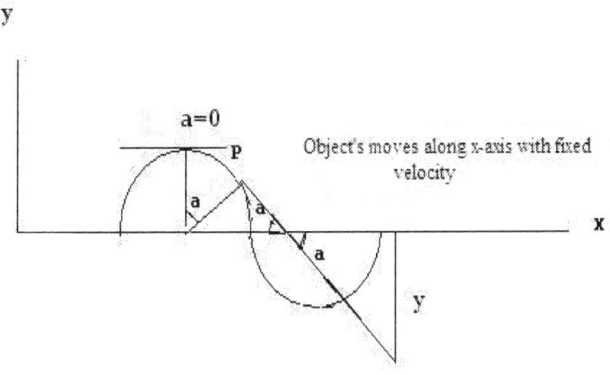

Particle's Speed

Thus as particle gets close to x-axis, somewhere along the line, it reaches the light speed (c = 300.000 km/s) and has to disappear from space-time (because according to Einstein's postulate, universe cannot accommodate speeds more that c). We have speculated that the particle has to exit space-time and enter singularity. In addition, somewhere below x-axis the particle will reappear because its speed decreases to c again and then reduces further to its ordinary propagation speed.

A plane wave equation can be written as:

$$\Psi(x_a) = e^{-ip_a x^a/h}$$

Where p is momentum. Adding $2\pi\hbar$ to the $p_a x^a$ will not change its quantity. The equation has a time-like period of $2\pi\hbar/p_0$ and a space-like period of $2\pi\hbar/p_1$.

So we may conclude that at each period there are times that momentum is zero.

On the other hand p = mv. So, at zero momentum moments either m or v has to be zero.

175

$2\pi\hbar\ /p_l$ tells us that along x direction, momentum is also periodic. This means momentum appears and disappears during each period as well.

In addition, wave function is a complex function and has an imaginary (out of space-time component in it). In the same chapter, we also have speculated that the mechanics of particles in this movement mimics a bouncing ball.

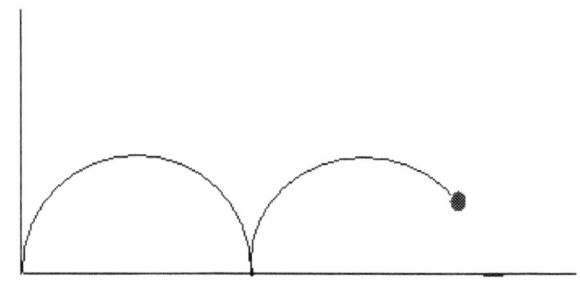

Bouncing Ball

Friction-less gravitational field

The two forces, which are acting on a bouncing ball, are attractive force of gravity and repulsive force of electromagnetic from molecules (electrons) at the surface of the earth. If the wave particle movement mimics the bouncing ball, in our model singularity has to have a repulsive force, which ejects the particles, and throw it to space-time. This gives the maximum kinetic energy at the time of entrance to space-time. For bouncing ball model to operate an opposite force is also needed. One can speculate that in wave particle scenario, the elasticity of displaced fabric of space provides the opposite force.

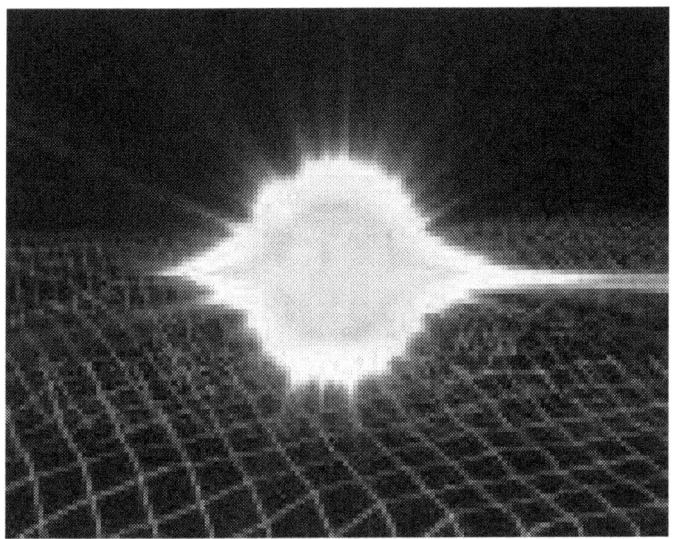

When the particle immerges into the space-time, it pushes the fabric of space and creates a bulge. As the energy of particle is used up, its speed decreases and particle reaches the peak of the wave. At this point, the elasticity of space-time pushes the particle down and back to singularity again.

Frequency in this Model

Please note that the frequency of each wave is proportionate to its magnitude square:

$$f \sim A^2$$

Where f is frequency and A is magnitude. This only sounds right. The more energy a particle has the higher it can jump and the bigger would be the distortion of space-time. In this scenario we do not have to assume pre-existing wave in fabric of space, rather we may hypothesize that if the amount of energy is harmonic with oscillation of space-time it creates the waves observed in micro scale. Any other energy level which is not harmonious with fabric of space either can not create wave or it will create short-life waves which belongs to unstable particles of standard model.

Therefore, unlike CIPA model, I am suggesting that energy belongs to object. The magnitude of object emergence inside space-time dictates its Compton Frequency (f $\sim A^2$) and therefore its identification.

Traveling back to singularity replenishes the energy of the particle and another cycle will start. This scenario is similar to water waves. Suppose a stone is thrown into a pond. The impact energy obtained by water molecules raises the wave but at the peak the gravity drags the molecules down back to the pond. There are no preexisting waves before the stone is dropped into the pond.

There are similarities between Sakharov-Puthoff model and the above scenario. In Sakharov-Puthoff model, the perturbation of particles in ZPE presence creates the mass and curvature of space. In the model in hand, the presence of matter (and its travel to singularity) is needed for transferring ZP energy to space-time.

In CIPA model, because the numbers of wave modes are enormous, and increases as the square of the frequency. The sum of tiny energy per mode times the huge spatial density of modes yields a very high energy density, which is not experienced. In our model, we confine the ZPE delivery just to individual particle-waves and not plane (waves that carry particles) and not all possible waves in microcosm. Therefore, this problem does not arise. This can only happen if ZPE exists outside and its energy being carried by particles into the space-time.

Preexisting Waves

Alternatively, we may postulate that plane waves introduced in Higg's Mechanism and QVIH may actually exist. The attractive and deflective force of zero point fluctuation may create randomly phased plane waves (shapes) in space-time.

Roger Penrose indicates that Einstein field equations:

"Predict a spectrum of quantum fluctuations in space-time. It also has a precise Planck-scale description, which makes use of very elegant mathematics connected to the invariants of graphs and knots."[5]

We may further assume that a particle according to its energy level gets involved with a harmonic wave-like curvature of space and continues its journey just like a planet or any other object traveling in a gravitational field or just like a car following the path of a curved road.

Bodies and Waves

The fundamental particles (fermions) have longer wavelengths. Therefore, in this scenario the waves in bigger scales belong to fundamental particles. Since the known particles are numbered and have specific frequencies, the micro–curvature of space cannot be completely random phased, rather the space waves created by ZPE at least in bigger wavelengths have to be numbered with specific frequencies. In addition, Pauli's Exclusion principle indicates that just a handful of useable waves in larger scales are formed. According to Pauli's principle particles occupy their territory exclusively. They do not share their trajectory with any other similar particles.

As we get to smaller wavelengths, we gradually get closer to the territory of hadrons and atoms and molecules and then larger objects. Bigger objects oscillate in smaller wavelengths. If you had a bowling ball with a mass, of say one kilogram, moving at one meter per second, its wavelength would be about a septillionth of a nanometer. This is so ridiculously small compared to the size of the bowling ball itself. That is why we never notice any wavelike motion while looking at a bigger object.

Here we can get philosophical and postulate that each one of us just like any other object have a wave like motion and enter and exit singularity in each period of our wave motion. Then this can explain some of the strange findings in transpersonal psychology experiments.

Mass and Wave Function

In the previous chapter, we presented the detailed format for particle-wave function in this model. Above, I also explained my conjecture about the nature of the rest mass of particles.

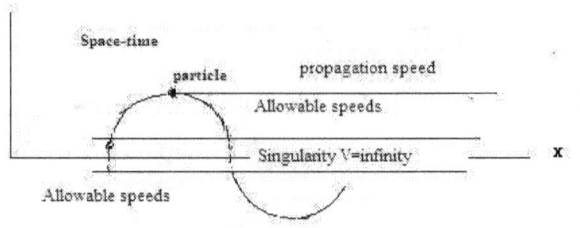

Modified C Diagram

In following paragraphs, I apply the above format to actual particles to see if the above model is in line with observations and experiments.

Fundamental Constants

Previously, I have mentioned that this chapter is the center-piece of this model. Here I offer explanations for the nature of Planck Constant, hierarchy problem and masses of fundamental particles as we go along.

First, let us discuss the fundamental constants. There are some numbers which frequently being encountered in mathematical calculations related to different astrophysical and quantum mechanical experiments in laboratories. We do not know where they come from. It seems that they are natural and originated from fundamentals that created our universe. That is why they are called Fundamental constants.

The most famous one is the light speed denoted by (c). Nobody knows why light constantly travels at approximately 300.000 km/s. Another major one is Planck constant (denoted as *h*) with the value of

$6.626\ 0693 \times 10^{-34}$ J s.

There are also others, like Gravitational constant, Boltzmann constant, Dirac's constant ($h/2\pi$) and the Columb Force constant. Masses of fundamental particles are also considered constant because apparently we do not know where their value comes from.

Unlike CIPA model, in this model energy does not belong to waves in the field. If it did, such energetic waves had to interfere with propagation of testable rays in space and would show their signatures in experiments. Such an effect has not been observed.

I believe energy accompanies the particles, which are traveling along the wave. I will take the lead from CIPA researchers postulate that,

"The energy of the ZPF continues to rise sharply with the frequency of the radiation quantitatively, the energy density is proportional to the cube of the frequency; double the frequency, and the energy increases by a factor of eight. At what frequency the ZPF spectrum finally cuts off or loses its ability to interact with matter are important and still unresolved issues?"[65]

What if particles scoop up energy from singularity twice during each period of their wavelength? Can this be the reason for energy to be proportionate to frequency of objects? Let's review it.

For the wavelength of a particle, we may write:

$$\lambda_X = h \,/\, m_X C$$

Where λ is the wavelength of the object, h is Planck constant, C is the speed of light, and m is the mass of the particle. Because, $\lambda_X = C/f_x$, we may write:[40]

$$C/f_x = h \,/\, m_X C$$
$$f_x = m_X C^2 /\, h$$

So, we conclude that the frequency of any object is proportionately related to its mass. Having the above scenario in mind, how do we explain the relationship?

Here we may postulate that, the mass is the kinetic energy obtained from singularity and delivered by an object each time it appears in space-time. Because in above model particle hits singularity twice in each wavelength, so the energy delivered is twice its frequency (f).

$E_k = E_s \, 2f$ (1)

Where E_k is the total kinetic energy of the particle, E_s is the unit of kinetic energy delivered in each trespassing. So objects with higher frequency deliver more kinetic energy and thus they are more massive.

To calculate E_s of different particles we used the Einstein equation, $E_k = m_0 c^2$. Where m_0 is the rest mass of the particle and c is the light speed. Then we write the total energy as product of Compton frequency of the particle and E_s;

$Ee = E_s * f_{com}$

Thus the energy derived from singularity is calculated as:

$E_s = * f_{com} / Ee$

Below, we calculated the E_s for electron and muon as examples. Please note that values for Compton frequency and mass of the particles are obtained by experiments in Fermi lab and elsewhere.

$Ee = m_0 c^2 = 9.109826 * 10^{-31} * 9 * 10^{16} = 81.988434 * 10^{-15}$
$Ee = E_s * f_{com}$
$f_{com} = c/\lambda_{com} = 3 * 10^8/2.426310215 * 10^{-12} = 1.236445357 * 10^{20}$
$E_s = Ee/ f_{com} = 81.988434 * 10^{-15}/1.236445357 * 10^{20}$
$= 66.30979164 * 10^{-35}$

The Es for Muon is:
$E_{mu} = m_0 c^2 = 1.88353140 * 10^{-28} * 9 * 10^{16} = 16.9517826 * 10^{-12}$
$f_{com} = c/\lambda_{com} = 3 * 10^8/11.73444197 * 10^{-15} = 0.255657662 * 10^{-23}$
$E_s = E_{mu}/ f_{com} = 16.9517826 * 10^{-12}/0.255657662 * 10^{-23}$
$= 66.30656976 * 10^{-35}$
The value for the Planck Constant is:
$h = 6.626\ 0693 * 10\ -35$ J.s

According to above calculations Es of muon equals the Es of electron and they both are equal to the Planck constant with a small margin of error.

The SI unit version of equation (1) can be written as:

$E_s = E_k/2f$, therefore E_s =joule/Hertz

This is similar to the SI unit of Planck Constant which is joule/Hertz[40]

We can do similar calculations for proton, neutron, tau and other objects.

Tau; $\lambda = 0.69770 \times 10^{-15}$
$m = 3.16788 \times 10^{-27}$ kg
E_s tau = 0.6630689628

Proton; $\lambda = 1.321409847 \times 10^{-15}$
$M = 1.67262158 \times 10^{-27}$ kg
Es proton = 0.66397957266952337X10-35

Neutron; $\lambda = 1.319590898 \times 10^{-15}$
$m = 1.67492716 \times 10^{-27}$ kg
E_s neutron = 0.663065590544696050X 10 -35
E_s of all of the above examples are equal to the Planck Constant.
So Es of different particles are equal and is equivalent to the Planck constant.

Assumption MG #1; Planck constant is the amount of energy delivered by particles to space-time in each period.
This is a serious speculation and the key possibility that our model for Particle–Wave Function and the principles of the presented model is valid.

Hierarchy Problem
Hierarchy problem occurs when the value of similar parameters obtained by experiment are vastly different and unrelated to each other. For example the ratio of mass of muon / mass of electron is:

$1.88353140 *10^{-28}/9.109826*10^{-34} = 206758.23$

One expects that masses of muon and electron which are both leptons being in similar order. But the mass of muon is 206758.23

the times Mass of electron. Similar irregularities exist between masses of other particles of the Standard Model. While the mass of up-quark is 0.004 GeV/c^2 the mass of top-quark is 176 GeV/c^2

If we take the Higgs waves and bypass Higgs boson then hierarchy problem is somewhat out of our way.

In this view particles get different masses according to the frequency of the wave they choose. A low frequency wave creates up quark with a very small mass.

Using the above formula we can see that the rest mass of particles is not arbitrary and can be calculated by product of energy from singularity (Planck Constant) and its frequency. So in this model, the other natural constant (Masses of fundamental particles) is also explained.

Assumption MG #2; Mass is the product of Planck constant and Compton frequency of the object.

$$M = f_c * h$$

This is consistent with the Compton wavelength equation: $m = h/\lambda_c c$ where λ_c is Compton wavelength and c is the speed of light. Therefore, I conclude that mass is the total energy delivered to space-time by an object.

The Thing

It seems that the particles are essentially made of one entity (a thing). The energy and therefore the path chosen by that thing (wavelengths) specify the identity of it and this is how we differentiate and name it as different fundamental particle. In beta decay (transformation of neutron to proton) a *d* quark changes to a *u* quark and releases a W boson which means releasing energy.

In the above model, we may postulate that in beta decay, a *d* quark with higher frequency looses some of it kinetic energy and therefore moves to another path with longer wavelength. This is when we call it *u* quark.

This can be an explanation for particles changing to each other. This is observed in accelerators everyday. Therefore we may speculate that different particles are actually the same thing. We distinguish and differentiate the thing by measuring the amount of kinetic energy and the path it follows (Compton wavelength).

Assumption MG #3; Fundamental particles are essentially the same entity except that their kinetic energy and therefore the wavelength adopted will differentiate them from each other.

Compton Wavelength

How would a particle choose which wave to follow? De Broglie introduced the Compton frequency as an intrinsic character of each particle or mass.

The Compton wavelength of a particle x is obtained by $\lambda_X = h / m_X C$ and Compton frequency by $f_X = m_X C/h$ where m_X is the mass of particle and h is the Planck constant.

Everything else being constant, mass is directly proportionate to frequency.

Previously, I have assumed that particles in their wave like motion enter and exit singularity relative to their frequency. We may assume that from here kinetic and potential energy will cause the wave motion to continue as long as they are not disturbed. Obviously as long as Compton wavelength of a particle does not change its character stays the same, But if Compton wavelength changes we are dealing with a new particle. Changing the Compton frequency is possible in high energy accelerators.

High Energy Accelerators

In accelerators the new particles are constantly formed and interchange to each other.

Inversely, we can postulate that the wave chosen depends on the amount of energy (kinetic energy) which is delivered in by a specific particle. Combining it with the assumption of preexisting plane waves we may conclude:

Assumption MG #4 Fundamental particles only associate with the waves which are harmonic to their Compton frequency.

Relativistic Mass

So far, we have been dealing with the mass of a particle at rest. Relativistic mass (the mass of an object that is traveling with high velocity in relation to an observer) is a different issue altogether.

When an observer is measuring a rest mass in his lab, since both observer and the mass are almost stationary to each other in the same frame of reference, the measured mass is rest mass (m_0). We may call the lab *frame of reference A*. But if the mass starts to move with velocity v, it is residing in a new frame of reference, we may call it *frame of reference B*. The two frames have a velocity relative to each other which is due to kinetic energy of frames of references. The Lorentz transformation equation for relativistic mass is as follows: $m = m_0 / \sqrt{1 - v^2/c^2}$

The equation suggests that as speed increases the mass increase as well.

How are we going to explain this difference in mass measurement?

Previously we have assumed that the nature of mass is kinetic energy. When the observer in frame A (a scientist in Kennedy Space Station) is going to measure the mass of an object which is moving in frame B (the flying shuttle), he has to take the extra kinetic energy of moving shuttle into his account. While moving, the inertia (the force needed to move the said object) has changed proportionate to the velocity (extra kinetic energy) of the shuttle.

If an astronaut inside the shuttle measures the same object, he will get the rest mass that is less than the mass that scientists on earth will record. The extra kinetic energy cannot be measured inside the shuttle, because the astronaut and the object are stationary relative to each other.

Dirac's Electron and this Model

In our interpretation mass is defined by kinetic energy. This kinetic energy is obtained from singularity. Clearly, there are some similarities between this model and Dirac's electron in Higgs mechanism although they are fundamentally different.

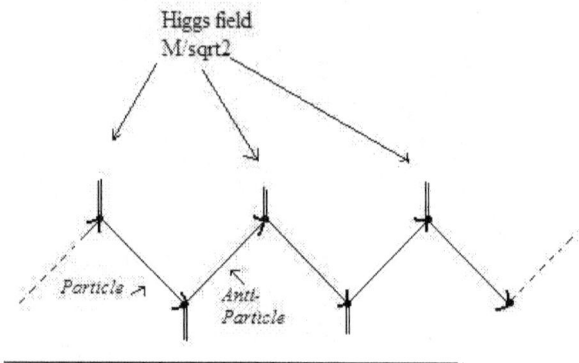

Dirac Electron and Higgs Field

Particles constantly convert to each other while droping and acquiring mass from
Higgs Field

Roger Penrose, Road to Reality[56]

The birth and rebirth of electron is in line with our model.
Also for a charged particle, the second portion of the phase (below
x-axis) has the opposite charge that acts as anti-particle.

On the other hand, the instantaneous speed of Dirac particles
is always constant and equal to speed of light. The variation in
propagation speed comes from zigzag motion of two components
as they average out. In our model, the speed along the propagation
line is constant but instantaneous speed of the particle changes
between propagation speed and speed of light in 2-dimension
wave plane. Vanishing and rebirth comes by entering and rising
from singularity.

Gravity in this Model

I undertake Sakharov-Puthoff model claims that the particle interaction with ZPE and perturbation of it is the origin of gravity and curving the space-time, as a lead to present an alternative model for gravity.

The Newton's law of universal gravitation conveys that massive objects attract each other. Newton's gravitational force is governed by the following equation:

$$F = G\, m_1 m_2 / r^2$$

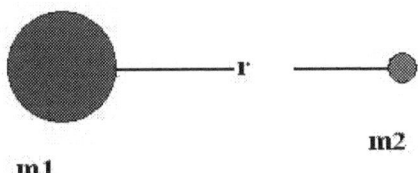

Newton's Gravitational force

Where F is the attractive force between two masses, m_1 and m_2. The distance between them is shown by r.

In contrast, Einstein's General Relativity denotes that a massive object curves the space-time and the free falling objects follow the curved space-time.

Mark McCutcheon, in his book *The Final Theory*, raised a very valid question. The gravitational force is constantly at work. As such, its energy is being constantly consumed. Therefore, one will expect that the force diminishes and disappear by the passage of time. This is against observations. The earth's gravitational pull has kept the moon in its orbit for more than four billion years. For the gravitational pull to stay unchanged, it has to replenish constantly. We need an unending source of energy to provide the gravity that exists throughout the universe.

The true nature of gravity is not yet understood but we have the Principle of Equivalence that tells us that the gravitational force corresponds to masses involved.

The Einstein field equations imply that any accelerated mass radiates energy. Similarly, the Maxwell equations indicate that any accelerated charge radiates electromagnetic energy.

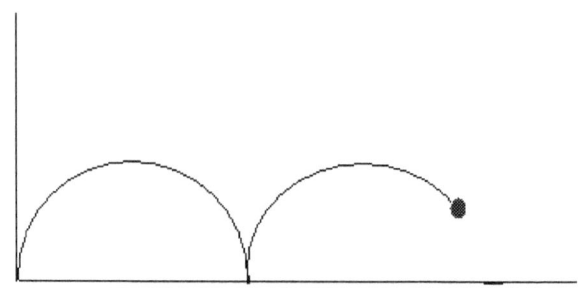

Bouncing Ball

Friction-less gravitational field

Previously, I have used a bouncing ball analogy for the particle-wave function. Please note that the velocity of a bouncing ball is always changing. In our model, particles are constantly accelerating or decelerating while following the path of their wavelength. Therefore, we can assume that the radiated energy exert the attractive force. We call this attractive force, gravity. This force is also proportionate to particles frequency and thus it's mass. This will satisfy the gravitational/mass equivalence principle.

In another scenario, if a particle strikes into space-time network, the network bulges in and its lines concave. The created dent depends on the elasticity of the network proper and the force applied. We can look at the sewing machine operation as an analogy. Frequent penetration of the needle of sewing machine creates a depression or dent in the fabric being stitched. If the frequency is high enough the concavity remains steady. This is an analogy for how a colliding particle can create curves similar to gravitational fields in space-time.

Please refer to the Wave-Particle Function chapter to see how, in this model, a particle is assumed to smack into the space-time network and create gravitational fields during its wave function.

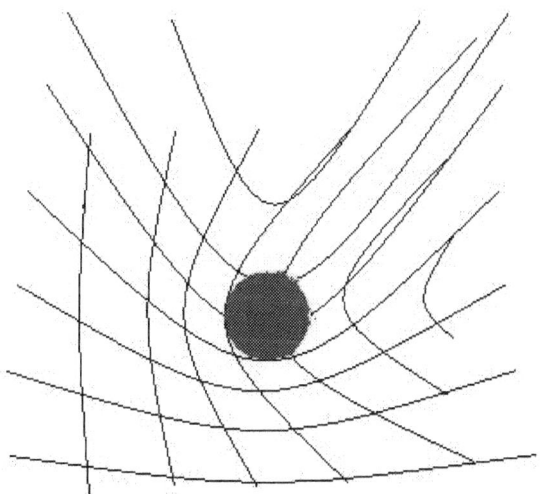

Object Impact on a 2-dimensional Space network (mesh) induced curves satisfies inverse distant square law

In this model, the long sought after and never found graviton (assumed responsible for the gravity effect) is not needed to create the gravitational curvature of space. The General Relativity theory asserts that any object in the gravitational field follows a straight-line trajectory in space. Gravitation happens because the space-time itself is curved by the presence of the mass.

The graviton does not seem to be necessary to create the gravitational effect. Rather the wave function of a massive object (by either radiating energy or mechanically striking the fabric of space-time) creates the curvature needed for the gravity effect in Einstein's theory of General Relativity. The frequent penetration of space-time by a massive object also creates the gravitational wave, which spreads throughout the universe.

Equivalence Principle

Looking at presented model for wave-particle, we find a good explanation for the equivalence of gravitational and inertial mass.

The Compton frequency is proportionate to the mass of the object. The energy of the object is also directly related to its frequency and the total mass that it caries. A wave function as described above would deliver a force to the fabric of space-time that is proportional to its mass and energy.

In the sewing machine analogy, the force on the needle creates the concavity and frequency of penetration if high enough can hold the concavity in place. We will have a steady depression if frequency 0vercomes the elasticity of the fabric.

Therefore, the mass of an object is directly related to the curvature produced in the space-time and as a result to the gravitational field.

In reality, the gravity is not only related to the mass of an object. More precisely, the gravity is related to the mass and the energy carried by it. For example, a compressed spring has more energy and therefore, creates more gravity.

The model presented above explains the above phenomenon. In our model, the curvature created is related to the total force inserted at the time of smashing into the fabric not just to the mass of the object.

Quantum Mechanics

Flourishing fountains fanfare out of a still pond

Quantum Mechanics is the science that studies the sub-atomic particles and deals with the behavior objects within their Compton wavelength.

While studying quantum mechanics, there are four points to keep in mind.

First, more than eighty years of lab experiments proved that the principles of quantum mechanics are real and valid. Mathematics of quantum theory precisely predicts and reproduces the results of physical experiments. We are applying these principles in making delicate instruments and building precise computers. Therefore, Quantum Mechanics is the actual science of subatomic arena.

Second, Quantum mechanics is weird. In this arena, certainty of classic physics is replaced by the uncertainty and super position of states. Casualty (cause and effect) may reverse so that cause appears before effect (second Feynman diagram). Direction of time can reverse also and future appears before the past. Entanglement between distant particles exists without apparent link. The law of conservation of energy gets pale in many instances. And so forth and so on.

193

Third, apparently we cannot utilize the logic of classical physics to interpret the weirdness of quantum mechanical phenomena. We have to develop a new logic to accommodate a reality that is beyond our today's consensus reality.

Fourth, quantum is not an abstract object that belongs to laboratories and scientists, it is the basic building block of our body and our macro-world and the laws of classical physics are constructed and originated from its mechanics.

One of the main tasks of this model is to propose explanations for unexplained findings in quantum theory. We cannot explain the Schrodinger's cat being alive and dead at the same time, with our conventional logic. Surprisingly though, we can imagine such a super position in our mind. Not only our imagination can accommodate such a dual and antagonistic state but our mind activity is frequently utilizes similar super positions in its different functions.

Niels Bohr's approach to Quantum Mechanics was just to observe the observations and not trying to find causality for their presence or explain the reality and history behind them. Albert Einstein on the other hand believed that universe is made of real objects with real and persistent properties. He believed that Quantum Mechanic theory is deficient and cannot reveal the hidden variables that create strange findings in sub-atomic arena. He further believed that these variables do exist. Einstein believed a deeper theory would find these variables, which are hidden in our experiments. Most physicists have not favored hidden variable theories. Experiments and calculation results contradict these theories. Although Bohemian mechanics (refer to David Bohem) tries to offer an explanation for it.[35]

Here I am introducing a non-local media, which is connecting different points of space together. Let's see if this model can offer reasonable explanations for different quantum mechanical paradoxes. I hope this model prove to be the Einstein's deeper theory, which explains the quantum mechanical experiments and brings the deterministic reality back to the picture. Furthermore, because this model has a mind component, it contains some of the interpretations and views of Neils Bohr as well.

Quantum Mechanic Domain

A particle generally behaves quantum mechanically when observed at distances shorter than its Compton wavelength. Therefore, we can conclude that there is a cut off between classical physics arena and quantum mechanics domain. Please note that we are not choosing *below certain distance* as a cutoff for quantum mechanical arena. The cut off line is at the edge of the Compton wavelength of any particle. Moreover, wavelengths of different particles are poles apart.

The Compton wavelength (λ) of a particle is given by;

$\lambda = h / m\ c,$ where h is the Planck Constant, m is the mass of particle and c is the speed of light.

What is with the mystery with the Compton wavelength? It seems something unfamiliar to classic physics is happening inside the wavelength of any particle that is responsible for weird phenomena that we are observing in quantum mechanics. I have explained my conjectures in the wave-particle function and other previous chapters.

Momentum

In classical physics, momentum is defined as the product of mass (m) and velocity (v) of an object.

$$P = mv$$

Simply speaking, momentum is the impact felt by a boxer receiving the opponent's tossed feast.

The relation between the momentums of an object in regards to its spatial position (x) are obtained by:

$$P_a = \alpha / \alpha x^a$$

Where α is a constant. If our object is a subatomic particle then we need to add imaginary number (i) and Dirac constant ($\hbar$) to the equation,

$$P_a = i\hbar\ \alpha / \alpha x^a$$

The Dirac constant is a reduced Planck Constant ($h/2\pi$).

195

The presence of *i* indicates that the momentum of a subatomic particle is governed by a complex function. Therefore, the momentum of subatomic particles is periodic (see the complex number chapter). Therefore, the Assertion C2 specifies that the value of the momentum has to hit zero at each period.

The presence of ℏ also points out that momentum is directly related to Planck constant. In wave-function chapter, we have assumed that the particle itself disappears and reappears in space-time in each Compton wavelength. We can relate the intermittent blinking of momentum to the intermittent emerging of the particle into the space-time.

Moreover, in the Mass & Gravity chapter we have assumed that the Planck Constant is the amount of kinetic energy carried by the particle upon its arrival into the space-time.

In following paragraphs, we will review the quantum mechanical phenomena while keeping the above conjectures in mind.

Heisenberg Uncertainty Principle

According to Werner K. Heisenberg, the famous German physicist, we cannot simultaneously determine the position and momentum of a particle at ultra short distances. This kind of correlation between two properties is called a complimentarity relation. The equation is written as,

$$\Delta E \, \Delta t \geq h/ 2\pi$$

In Heisenberg's famous uncertainty relation for position and momentum, when the position uncertainty changes in position (Δx) is less than the Compton wavelength, the momentum uncertainty changes in momentum (Δp) is greater than $h/ 2\pi$. Since momentum carries energy, the uncertainty in energy is greater than, $h/ 2\pi$. This implies that if we pin point a particle's location, its momentum can vary widely and therefore we can not be certain about its momentum.

To justify the uncertainty principle, error in measurement and lack of appropriate tool are brought up. Neither is very convincing.

Therefore, quantum theory tells us that we cannot track a subatomic particle by any method whatsoever. Can we assume that, we cannot detect particles because they loose their mass and leave the space-time? Maybe we have to change the sentence as "we cannot track a subatomic particle by any method whatsoever in *objective world*. The problem arises when we are expecting to see the whole picture in just one arena, (the real number arena). As an analogy, please note that we cannot follow and understand a three dimensional motion in it's entirely in a two dimensional world. Evan Walker says: "... Heisenberg used matrices (whole array of numbers) to represent the positions and motions of an atomic particle.[8]

In his calculation to create the matrix he used the symbol *i* which stands for square root of −1, the so called, imaginary number. He had to choose a number, which is out of the domain of our real number system. We cannot ignore the quantity *i* and call it imaginary. We have to accept that it stands for a kind of reality. According to Dr. Walker quantum mathematics drags us to a scope "that is really an infinity of imaginary worlds."[8]

Therefore, we have to expand the science domain to include worlds other than familiar space-time. If mathematics so precisely is predicting the mystery world of quantum behavior, we have to value its elements. We have to accept that its unexplained or inapplicable measures to our physical world have an actual meaning.

Let us revisit the location /momentum uncertainty. Please note that one of the elements is spatial and the other is energy related. We can interpret the principle as; when locality gets blurred the energy is more defined. This was explained in the Boundaries Chapter. We also can extend the Heisenberg principle to time and energy in a system.

$$\Delta E\ \Delta t \geq h/\ 2\pi$$

For example a radioactive atomic nucleus decays with time. If the lifetime of such a state is Δt, then the energy of the exited states is uncertain by:

$$\Delta E \geq \hbar/2\pi\ \Delta t$$

197

The second diagram of Feynman for Compton Scattering also leads us to above uncertainty. Here again while *t* is a spatial element, the other element is energy. As time gets hazy energy gets more distinct and vise versa.

Spectral/Spatial Complemetarity

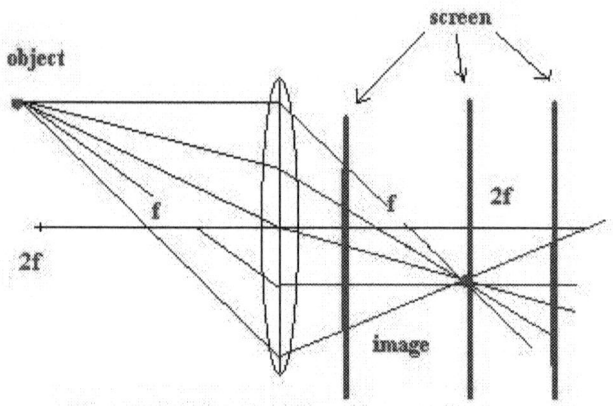

Spectral / Spatial Complemetarity
Relation

We can also extend the uncertainty principle to information carried in a beam of light after it passes a lens. Depending on the location of a screen behind the lens, we can either have a sharp image of the object in experiment, or we may come up with a fuzzy image or no image at all. On a proper location every portion of image data has a defined spatial location. Whereas in no image locations the information is carried by light rays are at spectral version and they do not exist in spatial location. Interestingly, in spectral form the data are non local. Which means every small portion of space carries all of the information about our object in hand.

Here we have a complimentarity relation between spectral and spatial features. As we come close to focal point we depart from spectral phase and come to locality. With the same token when we leave focal point we enter spectral phase again and the non-locality prevails. Isn't this a good example for understanding the

198

relation between a spatial local world and non-spatial non-local energy-informatics domain? In our model this paradigm is the proposed singularity.

Information/ Interference Compementarity

George Greenstein and Arthur Zajong mention another very interesting complimentarity relation in their book "The Quantum Challenge". They explained the complimentarity between information and interference pattern in double slit experiment (see the explanation under the same subtitle in this chapter). Referring to moveable slit modification of the experiment, they wrote:

"Wootters and Zorek have returned to Einstein's modification of the classic double-slit interference experiment, and analyzed it from the stand point of partial information...The slits are free to move. After particle has past through, we measure the slit's momentum... if the slits are moving downward, the particle must have past through slit 2...

Wootters and Zorek noted that the above conclusion is not in fact entirely certain. The same motion of the slits would also be observed had the particle passed the wrong slit --if the initial slit momentum had been large and downward...

They evaluated the probability of the initial slit momentum being large enough to yield such an erroneous conclusion, and so obtained an expression of the probability that we had obtained path information. Using the same wave function, they also calculated the resulting pattern of arrival at the final screen. It turned out to be a partially smeared-out interference pattern"[11]

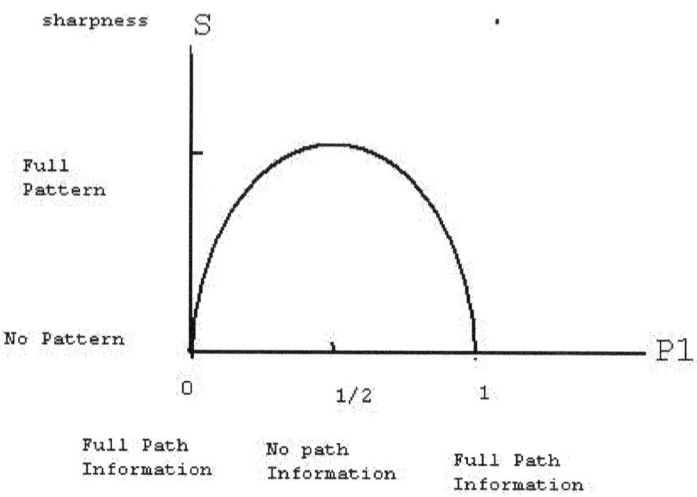

Reproduced from Reference #11

Their experiment showed that if they had certain knowledge about the slit that particle passed through, there where no interference on screen. But as the uncertainty about the slit increased a better interference were developed.

Looking at different complimentary pairs, the elements for each uncertainty relation are different in that, one of them is either spatial or mass type, and the other is either energy type or informational type. The two characteristic that we hypothesized for proposed singularity. The fuzzy states of fundamental elements in boundaries of space-time were discussed in Boundaries Chapter.

200

Schrodinger's Cat: Quantum Super Position of States

*For details check the link: http://www.upscale.utoronto.ca/
GeneralInterest/Harrison/SchrodCat/SchrodCat.html*

While in macrocosm the objects are present in just one state, in ultra small scale the objects demonstrate super position of states. For example, in classic physics an object will rotate either clockwise or counterclockwise. In microcosm though, one particle is spinning in both directions simultaneously. The analogy of a cat being dead and alive at the same time after a bullet was shot towards her is famous and is named after Erwin Schrodinger who first explained the super-position of states in Quantum Mechanics.

The bullet both misses and hits the cat at the same time. The Schrodinger's cat is simultaneously dead and alive.   Such a sentence does not relay any meaningful concept to us. Or does it?

In the observable world a cat is either dead or alive. The series of events lead to just one of the possibilities.

A bullet is shot. Multi universe theories claim that there are different universes that accommodate different possible outcomes, of each action. This cannot be acceptable. There are countless actions in each miniscule of time in universe and far more possibilities. Possibilities are endless. It means we have to have endless numbers of universes and the number is growing every second at a rate that transcends all concepts of infinitudes. It is also grossly against conservation of energy law if we choose to hold it. This concept is not economical either.

On the other hand, Schrodinger's wave equation which represent the super-position of states, also contains the imaginary factor *i*. Thus quantum state "will always turn out to contain terms that are imaginary...The complex character of the wave function in Schrodinger's wave equation means that what is there in a sense hidden from us."[8] In other words, Somehow in quantum arena we are exposed to out of space-time realm.

Renormalization in Schrodingers's Equation

The Schrödinger's probability equation for the position of a particle during its wave function can be written as:

$\Psi(x, y, z, t)$ = Probability Amplitude
$\Psi^* \Psi$ = Probability

If we believe that the particle is somewhere in space but we cannot exactly pin point it, we can normalize the formula and write:

$$\int \Psi^* \Psi \, cr = 1$$

Which means that we have altered the formula in a way that would show the probability of the particle being somewhere in space is %100. We do this because we believe that everything is confined inside the space-time. So, it is natural to assume that the particle is in space somewhere. Therefore, we write a formula to reflect a normal situation by our space-time logic. If we are revisiting the existing concepts then we are allowed to question normalization as well. Normalization is a major parts of quantum mechanical calculations. If we take normalization out of quantum calculations, we tremble the pillars of its existing mathematics.

Questioning the normalization is similar to questioning the attempts to understand the reality on the basis of space-time properties alone. This is a courageous action. But haven't we been very bold so far? As a matter of fact, solving these great mysteries need enormous amount of daring. At the same time revisiting the calculations can confirm or reject the validity of the singularity concept presented above. It seems that if we redefine the particle wave function to include singularity, we will come to a logical explanation for the phenomena that we are facing. I do not see why we shouldn't let our imagination explore that possibility. Aren't we defining the location of each particle with a complex number, which contains imaginary part as inseparable feature?

Decoherence

Decoherence – the blossom of just one state in macrocosm out of infinite quantum superposition states in micro scale.

In Quantum Mechanics, the Decoherence Principle explains why we do not observe superposition of state in our everyday experience. If we did, there would be an in-deterministic and chaotic world that nobody could stand it. Nothing would remain to rely on or build into it. Even we ourselves would be a total mess. As mixed up as the state that our mind normally is. Decoherence principle simply points to the fact that each particle normally exist in an environment. This environment is always bombarded with other particles and photons. The interaction of particles with each other takes them out of super position of state and leaves them in a definite state. That is how the world comes out of confusion and we see one definite state around us. This is a world where we can live and be able to rely on.

Singularity, Super Position Domain

If we cannot normally observe the superposition, then where is the most probable place to comprehend it? Where can we find such a cat, which is in superposition of life and death? We assumed singularity to contain information of an object but it lacks matter. In such a domain, information about any possibility can exist simultaneously. However, to materialize, the information has to be reduced to one state in space-time so that it become objective.

Are we accustomed to such phenomena or is it completely strange to us? Superposition of states happens in quantum level. According to the concept presented, the physical world connects and intermingles with singularity in every miniscule of space. In addition, we have assumed quantum arena to be the interface between our physical world and singularity. Surprisingly, such a superposition can happen in our mind as well. We are familiar with this kind of superposition of states. In different mind activities such as predicting upcoming events, we imagine different states of proceedings in order to compare and evaluate. When we are planning to make a decision, we picture different mode of possibilities. When we make our decision materializes and the outcome will be just one state.

As a matter of fact, the Copenhagen interpretation of quantum mechanics as first introduced by Neils Bohr implies that superposition of states is a mathematical formalism and exists in the mind of observer. Physical world is the reduced state of superposition which exists in our mind. So we reduce the super position to one possibility and somehow project it in physical world. See how all of these fall together as different pieces of the same puzzle. Please refer to the three circles of reality presented at the Introduction Chapter.

Singularity as an informational domain can contain a superposition of states just like our consciousness, which can hold such a superposition. When the particle enters the objective space-time, it is reduced to one state and becomes observable.

Defying the Conservation Law

As mentioned before, if we take superposition as actual objective phenomena, it turns out to be against the dearly loved conservation of energy/matter law. Many discussions and arguments are brought forward by trying to expand super position of states to macroscopic world. One of the main arguments is the many worlds of Everett presented at 1957. It essentially says that for any possible state of any quantum system a new universe will start which accommodates that particular state. Although this interpretation is revoltingly against conservation law and grossly non-economical and beyond belief, it has gained popularity.

In the Consistent History approach to quantum mechanics by Griffith, Omness and Gell-mann/Hartle the superposition state of quantum level is considered a coarse grained state while a projector reduces and refines it to just one state and project it to macroscopic level. According to them this is why we do not observe superposition in macro world. In 1995 Fay Dawker and Adrian Kent elaborated on the concept further. At least here reduction to one state is recognized while in many worlds of Everett reduction does not occur.

The super –position domain

The other dilemma is when and where and in what level this transformation to reduced state is happening. Please note that mathematics of superposition is continuous where as reduction represents a discontinuity and change in state vector.[56]

So the coarse grain (superposition) has to be in a continuous media where as the projected reduced state has to be in a discrete background. In proposed model space-time is discrete and elements in proposed singularity have a continuous nature. So again in my interpretation, the course grain state (superposition) happens in singularity and exists at informational level. The refined and reduced state is projected and materialized to our space-time. So boundaries of space time are where transformation actually happens. This is in line with the Copenhagen Interpretation of Quantum Mechanics as originally suggested by Neils Bohr. He believed superposition takes place at the mind of observer in informational state. Reduction occurs when we refine the information.

Do we need to extend the superposition to macroscopic world and then fall into the doldrums of infinite number universes, which are growing enormously by instant? Even thinking about it makes one insane. My humble suggestion is to consider the super position presence somehow out of our universe. Let us then take the state reduction as an event inside our physical world. This way we are left only with one universe. This is in line with our observation.

Here are the main questions: Are we going to take the every moment experience with mind behavior as a main portion of reality and accepting it as model to clarify the obscure portions of physical world?

Or do we keep trying to include implausible fantasies, like many universes or multiple histories (the idea that every event that could happen in the past actually happened) as part of reality? What I am suggesting in here is to find a solution by simply opening out the scope and include a mind like entity as a contributing domain. Just like mathematicians who when they could not solve problems in existing domain, opened out their scope and added negative numbers or imaginary number domains and opened the gridlock.

Tonomura Double Slit Experiment:
Feynman's *Sum over Paths Approach*

In Thomas Young's double-slit experiment a beam of light is directed towards a barrier with two slits that diffract the beam. A screen is installed behind the barrier that shows light and dark rows or the so-called interference pattern.   This is the basic experiment, which shows the wave property of light.   In 1920, Albert Einstein received the Nobel Prize for introducing the photon as a packet of light energy. Thereafter, light has been considered as a particle wave function. Photon is the particle portion of light.

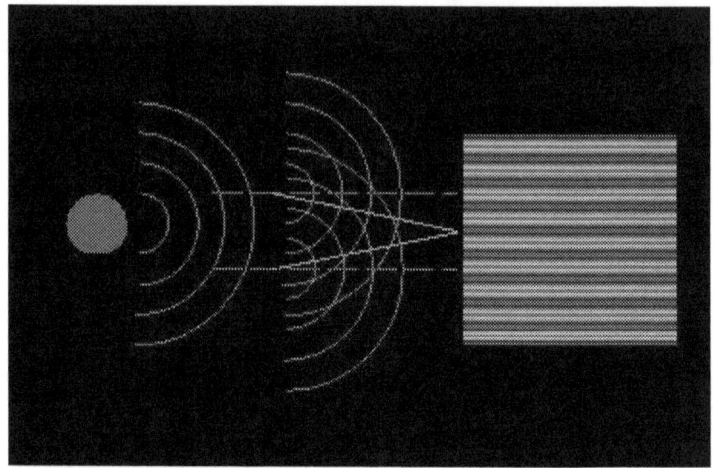

Tonomura Double slit experiment is almost a similar experiment but with a very strange result. In this experiment, electrons are fired one by one in time intervals of ten seconds. Interestingly, the interference pattern appears in the screen similar to the one when a bunch of electrons is fired towards both slits simultaneously. The experiment has been repeated by many researchers. Brian Greene explains how the strange results were first observed:

"In 1920 Davisson and Germer ... were studying how a beam of electron bounces off a chunk of nickel. The nickel crystals in such an experiment act very much like the two slits in the double slits experiment of Thomas Young... Their experiment therefore showed that electrons exhibit interference phenomena... even if the beam of fired electrons was thinned so that, for instance, only one electron was emitted every ten second, the individual electron still built up the bright and dark bands."[1]

In order to explain the electron two-slit paradox the Late Physicist Richard Feynman proclaimed:

"Each electron that makes it through to the phosphorescent screen actually goes through both slits. Feynman argued in traveling from the source to a given point on the phosphorescent screen each individual electron actually traverse every possible trajectory simultaneously...It goes in a nice orderly way through the left slit. It simultaneously also goes in a nice orderly way through the right slit. It heads toward the left slit, but suddenly changes courses and heads through the right. It meanders back and forth, finally passing through the left slit. It goes on a long journey to Andromeda galaxy before turning back and passing through the left slit on its way to the screen. And on it goes- the electron, according to Feynman, simultaneously sniffs out every possible path connecting its starting location with its final destination."[1]

Of course, this is not his personal opinion. Different experiments suggest the above conclusion. The so-called Diffraction Grating Mirror works with the same principle.

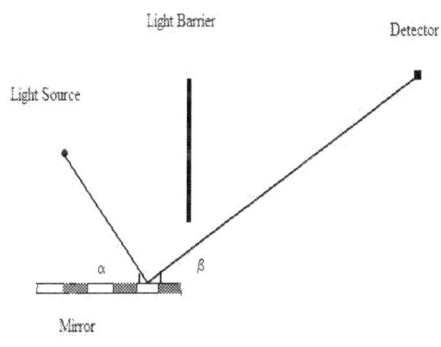

Diffraction Grating Mirror

Borken Mirror. The reflection angles do not match, but a detector placed out of range can still register a signal.

Broken Mirror
The reflection angles do not match. However, a detector placed out of range can still register a signal.[52]

In addition, the above explanation is obviously against the Special Relativity that limits the velocity to light speed. If the electron is going back and forth in different ways, it should have infinite fold time speed of light, which is contradicting the known space-time physics.

Singularity and Tonomura Double Slit Experiment

To have an infinite speed, the particle has to perform in a non-local arena. A better explanation can be reached, if we accept that the electrons are acting in a field with no time dimension. In this situation, speed conflict is solved but the paradox is still not completely resolved. One has to delete the notion of space and distance out of this experiment as well. This way the electron can come from everywhere and nowhere. Using the verb *come* applies to travel in space. It is better to use *appear* and accept that in this experiment electrons pop in and out of space-time universe.

On above interference experiment, Feynman tried to explain the paradox by claiming that each electron will follow all possible tracks before it hits the screen. We can claim that here, the electron follows the non-locality characteristic of the proposed singularity. Either non-locality or being not time-bound can explain this effect more logically than assuming that electron traveled along all possible trajectories before hitting the screen. Roger Penrose offers the mathematic equation for two-slit experiment by using complex numbers for quantum state. Roger Penrose writes:

"They can be represented on a two-dimensional plot with the purely real numbers running along the x-axis, and the purely imaginary numbers running up the y-axis, the imaginary axis."[5]

To introduce the mathematics Roger Penrose has to use imaginary dimension along real dimension. He could not find any real dimension in our space-time universe, which can help him to mathematically explain the phenomenon. Again, we can imagine the complex numbers in our mind. But quantum two-slit behavior is happening all over the world in every moment, even if, our mind is not with it. There should be another being out there to accept the image, to be able to accommodate the imaginary dimension of complex numbers. Can we suppose that singularity is there to do it for us?

One can assume that if complex numbers represent particle functions, then particles themselves have to have an imaginary (out of space-time) phase.

Abstract World of Mathematics

Roger Penrose uses the Argand Diagram (presented in the Complex Number chapter) to discuss the complex numbers behavior.

"The fact that these numbers are built into the foundation of quantum theory often makes people feel that theory is a rather abstract and unknowable kind of thing, but once you get used to complex numbers, particularly after playing around with them on the Argand diagram, they become very concrete objects and you do not worry so much about them."[5]

Therefore, Professor Penrose is suggesting staying in the abstract world of mathematics and viewing the imaginary number as a component of Argand diagram which is drawn in front of us.

If we try to understand the meaning of complex numbers with our space-time knowledge, we will face a concrete wall. Therefore, we choose to take refuge in mathematical domain. We tend to take imaginary number as a mathematical entity and give ourselves comfort. From school times we have learned how to use numbers as abstract entities. We are used to perform mathematical calculations and ignore their relation with objective reality. We cannot stay in this refuge for long. Our father Adam didn't. We cannot do it either. We have to explore, find and assign a tangible and deterministic reality to imaginary and notion of complex numbers. On the other hand because mathematics are so precise in calculating the Newtonian physics, electromagnetism, GTR and quantum mechanics, if in some parts it does not match our perception of reality, one could suspect that our perception might be wrong or incomplete.

In our model, because, there is no dimension in singularity an electron can move radially -- in and out of space-time universe, but yet cover the whole angular range, because singularity is everywhere and nowhere. An electron being at the same time in singularity, can defy 'steel ball' or 'Classical' interpretations of stationary orbital as perpetual motion in our space-time universe. In this explanation an electron does not need to follow the laws of classical physics.

Quantum Entanglement

Different Quantum Mechanical experiments have proved that pair of particles, which traveling back to back in opposite direction even if they are worlds apart, are entangled. Any change implicated to any of them will instantly change the other particle accordingly. It is interesting that the change occurs instantly.

This is one of the discrepancies between the Special Theory of Relativity and Quantum Mechanics. According to Special Relativity, nothing can travel faster than light. Instant connection between two spatially separate objects observed in Quantum Mechanics, is in contrast with the principles of the Special Relativity.

Professor Roger Penrose describes the entanglement as:

"Modern preoccupation with action-at-a-distance (to which Einstein objected) revolves around a purely quantum phenomenon. This is the celebrated thought experiment in which an 'entangled' system of two objects is created. The wave function of object # 1 is correlated with the wave function of object # 2. By the axioms of quantum mechanics a measurement of the properties of object #1 forces its wave function to collapse, instantly implies a correlated collapse in object #2 though it be arbitrarily far away."[5]

Brent Nelson, M.A. Physics, Ph.D. Student, UC Berkeley writes: "Non-local effects...occur in quantum mechanics and they cannot be understood in terms of one thing being separate from another - some sort of global activity is taking place."

Scientists have been puzzled by this strange phenomenon. Action at the distance without any apparent link also exists in electromagnetic field or gravity field. In case of quantum entanglement, no logical explanation, which can be agreeable by all, has been offered so far.

Aspect's Experiment

Alain Aspect and colleagues in 1982 experimentally proved the reality of Quantum Entanglement, and the non-local actions. The diagram below sketches the experiment for more detail, please refer to http://roxanne.roxanne.org/epr/index.html or other related academic books and resources.

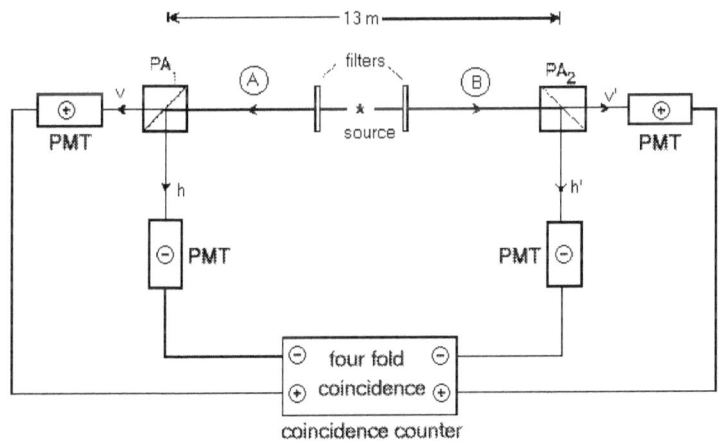

Aspect's experiment for quantum entanglement[25]

Allan Aspect designed and performed the above experiment to examine the concept of quantum entanglement.

A similar experiment has been repeated in different parts of the world since then. They all proved the entanglement exists between pair of particles that are far apart in space. Albert Einstein struggling to stick to a deterministic world suggested that there are contributing variables at work that produces this effect. He believed these variables are hidden from us because of our limited knowledge. However, the hidden variable explanation for the entanglement has been proved wrong by different experiments. Entanglement exists without any space-time link.

Singularity: The Media for Quantum Entanglement

If by our assumption, singularity collects remote distances of our universe in one spot (singularity itself) action at a distance can be simply described. Look at an analogy of image of a landscape in a two-dimension mirror. The images of the objects, which are far apart in three-dimensional world, but are in a line perpendicular to surface of the mirror, are superimposed and in contact with each other in two-dimentiality of a mirror. As previously mentioned in no-dimension, all points will fall on top of each other. One may argue that the images fall on each other not real objects. The above statement is true, but image is carrying the information. Information is not separate from object itself. In one view, the object is information in its totality.

Besides, in the Wave-function Chapter, I have assumed that objects enter informational domain (singularity) during each wavelength. The particles can exchange information instantly in this realm. Remember, there is no time element in the proposed singularity.

Ervin László, the famous Hungarian philosopher of science, postulated a similar scenario. He assumed quantum vacuum as a universal field that interact with matter. He asserts that the field:

"…acts as a holographic medium, registering and conserving the scalar wave-transform of the 3-dimensional configuration spaces assumed by matter in space. This universal fifth field is not inferred from space-time interactions like gravitational, electromagnetic, the strong and weak nuclear forces. In this new type of field, space and time become implicate, enfolded, as described mathematically by Bohm. The fifth field is spectrally (holographically) organized, and is made of the energy present in the interference patterns of the waveforms. The transformations from space-time order to this spectrum dimension are described by holographic mathematical formulations."[20]

The idea of space-time being embedded in a sea of something is nothing new. Paul Dirac also speculated that space-time is embedded in a sea of photon. He later on postulated that universe is embedded in a sea of negative electrons. However, such a sea has to have zero dimensions to be able to incorporate quantum entanglement. Singularity/Space-time duality with proposed description for wave particle function presented in this book provides such a zero size media where quantum entanglement can materialize.

The description of entanglement in this model is compatible with the Special Relativity because it infers that although a pair of particles may be far apart in space-time, in singularity domain they are related. Therefore, information does not actually travel beyond speed of light in space-time to reach the other particle.

This is another indication that the presented concept deserves attention. The explanation for action-at-a-distance is embedded in this model. As far as I know, there is not many theories out there that offers an explanation for quantum entanglement phenomenon.

Quantum Tunneling

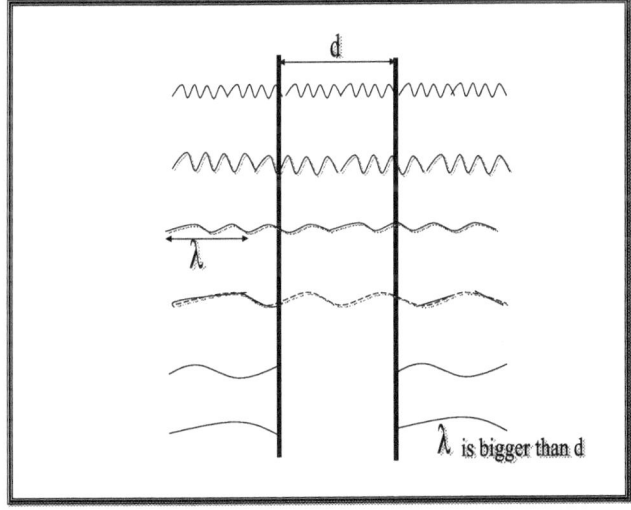

Cassimir Effect
Courtesy of Nahid Sahel Gozin

215

During its wave function, if a particle encounters a barrier where it cannot occupy an energy state (can not be present in the space-time), it tunnels and appears on the other side of the barrier. Quantum tunneling also does not have any explanation with Newtonian Physics. An explanation can be offered for the phenomenon that is similar the one that I offered for wave particle function. In quantum tunneling if condition does not allow, the trace of particle will not resurface and stays in singularity and shows up again when the obstacle in space-time is bypassed.

Super-Luminal Speed

Quantum entanglement and quantum teleportation are examples of the ability of saving information and playing it back as needed. In recent research by Dr. Mohammad Mojtahedi, of the University of New Mexico Physics Dept., pulses have been measured which travel faster than speed of light in a vacuum. He measured pulses up to 2.38 times the speed of light. Although different interpretations and conclusions can be derived from this experiment, he attributes results to quantum tunneling effect. We have supposed that, in tunneling the trace of particle does not appear in space in time and leaps to farther location. Therefore, quantum leap can be the reason that light's speed limit is violated in this experiment. One may assume, in quantum leap, when the particle resurfaces the collective speed is more than light speed in space-time.

Observer and Quantum State Reduction

As mentioned before, quantum mechanics tells us that in quantum level the particles exist in different states simultaneously (Schrödinger's cat). Remember that in quantum level the Schrodinger's cat is alive and dead at the same time. Experiments show that the act of measurement by the experimenter will reduce the state to one definite state. Is it the measuring or measurer who changes the overlapped different states to one single one?

According to Evan Walker the most accepted interpretation of quantum mechanics (Copenhagen interpretation) indicates:

"... that the system undergoes state vector collapse because of our mind. This effort to obtain an entirely practical interpretation of quantum mechanics ... lead us to the incredible conclusion that mind, or consciousness, affects matter."[8]

State reduction is not the only place where the measurer and the act of measuring influence the quantum mechanics. The observer also affects quantum entanglement. In entanglement two or more subatomic particles are connected to each other even if they are worlds apart. If the spin of one of them changes the spin of the other particles change accordingly and immediately. Considering the history of particles in the world logically each particle should have countless entanglements with countless particles which it encountered in its past history.

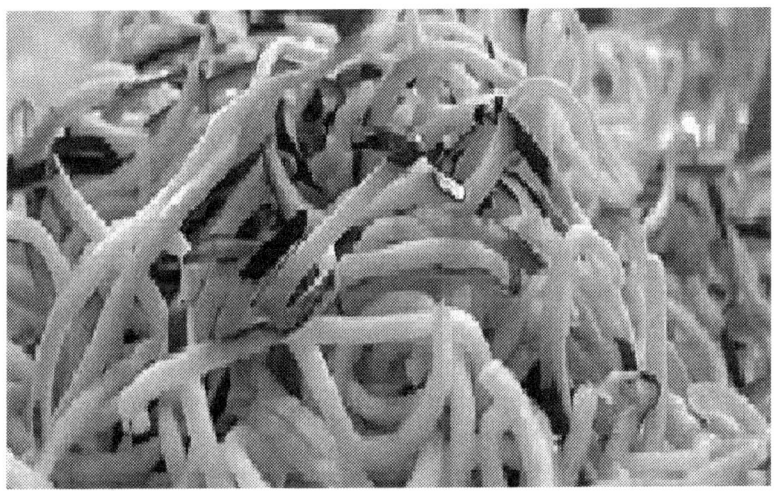

Surprisingly, when a physicist begins the entanglement experiment between a pair of particles any preexisting entanglement is eliminated and just the entanglement between two particles under experiment is observed. So we as experimenters are changing the world even at the quantum level.

Understanding how human can affect quantum mechanics is not easy. It has created many controversies. One wonders, if the principle can be applied in everyday life. Can we change the world at our will?

One school of thought believes that the history and the state of the world depend on our question. We see the answers, which are in line with what we are looking for. Others like Lee Smolin believe there is "one universe seen by many observers, rather than many universes, seen by one mythical observer outside the universe."[27]

State Reduction and Dream Elucidation

Copenhagen interpretation of quantum mechanics has gained popularity against the other theories despite its very strange sketch. It talks about how an observer would collapse superposition of particle properties into a definite and objective state. Arnold Mindell compares these phenomena with fantasy and its overlapped states versus a definite state found in consensus reality. He claims:

"Another method of marginalizing experience occurs when you start thinking about the meaning of the fantasy, while in the midst of the fantasizing. Thinking about it stops the fantasy."[14]

Singularity and State Reduction

Lee Smolin believes that we are seeing the same universe from different angles. At the beginning of this paper I was trying to draw an entity with zero dimension where the image of all of the points in our world can drop in it and stay superimposed. Some place where different events can happen simultaneously because the notion of time is not present. May be we don't have to speculate in remote and strange thought domains. Multiple universes and multiple histories do not sound very logical. Rather we can assume the intimate proximity and relationship between macrocosm and singularity. With this assumption, multiple states stay in one arena, which has the capability to accommodate them. This way we do not have to mix it up with activities, which are allowable in space-time and macrocosm. Otherwise, it leads to confusion and we will be dragged into an indeterministic physics that nobody knows how to deal with it.

Dr. Walker rejects the measurement error interpretation by reasoning that our measuring device is only another sub-atomic particle, which according to Schrodinger's equation entangles with the particles under observation. This should add to complexity of superposition, not state reduction. At the beginning we explored the similarities between mind and proposed singularity. The necessity to utilize complex numbers in order to explain quantum behavior suggests that elements in imaginary world participate in determining quantum physics. Previously, I took the imaginary portion, to represent the effect of the proposed singularity.

Copenhagen Interpretation

Copenhagen interpretation suggests that observer is changing the reality. In following sentences, Dr. Even Walker explores state reduction paradox.

"The answer would seem to be that something is wrong with quantum mechanics, that its make-up is incomplete, or that something about the act of conscious observation alters the quantum mechanical description of reality- something that plays a special role in quantum theory" [8]

This is in line with the psychological notion that our positive or negative thinking can change the outcome.

Telekinesis

Telekinesis claims that mind can move the objects in macroscopic world. The phenomenon is not scientifically confirmed yet. But its advocates claim that none of the four physical forces has been detected as a force behind the movement. Para-psychological activities are claimed and demonstrated, but we cannot explain them. Quantum behavior is repeatedly shown and recorded in numerous experiments and we are applying it to our advantage but we cannot explain it either. Consciousness is in us. It is the basis of our awareness, but still we cannot explain it. Certain basic mathematical functions and facts exist (like complex numbers) but they cannot be related to physical world in its generality. To understand the unexplained we have to be ready to look for new dimensions and domains. We have to be ready to travel to domains which goes beyond matter or other known physical properties, if needed. Pascual Jordan, one of the major early contributors to quantum theory says: "We compel [the electron] to assume a definite position." [8]

Who are we? Where do we stand in the equation? Can we come to the conclusion that our consciousness is the extension of singularity?

Can we claim that our will is creating the state vector collapse according to our desire?

Quantum Teleportation

We have watched the fantasy of the teleportation in the Star Trek series. The actual quantum teleportation is the transferring of tiny units of information, called quantum bits or qubits, about a particle. By transporting the information to another location, we can create a replica of the particle in new location. In quantum teleportation actual matter or energy are not transported.

This procedure was done inside laboratory in short distances. Samuel Braunstein, a professor of informatics at the University of Wales in Bangor, England was part of a team that teleported photons from one end of a table to the next in 1998. Nicolas Gisin, a physicist at the University of Geneva and his team teleported qubits carried by photons of 0.05 inch (1.3mm) wavelength in one laboratory onto photons of 0.06 inch (1.55mm) wavelength in another laboratory 180 feet (55 meters) away along 1.2 miles (2 kilometers) of fiber optic wire in January 2003. And most recently, Austrian researchers have teleported photons across the Danube River in Vienna. They reported their achievement in April 2004. Because of measurement problems created by uncertainty principle, teleportation in above experiments had to be performed by using entanglement principle of quantum theory. Below I am including one of the reports regarding this phenomenon provided by Daniel F. James from University of Innsbruck in Austria, in collaboration with a scientist at Los Alamos National Laboratory.

"In the experiment described in the scientific journal Nature, the group achieved teleportation using singly-ionized calcium atoms that were confined and cooled to ultra-low temperatures (around 15 millionths of a degree above absolute zero). Using lasers, the internal configurations of the atoms -- their quantum states -- were controlled very precisely, allowing entanglement between two of the atoms to be created. One of these entangled atoms was then further entangled with a third atom -- the input of the teleporter. By performing a simple measurement on this pair, and another series of interactions dependent on the outcome of the measurement, the original input state was then re-created on the remaining (output) atom. The quantum state teleportation experiments were carried out at the

University of Innsbruck's Institute for Experimental Physics."

Singularity and Teleportation

Please note that teleportation was performed by just transferring the information, not the actual particle. In other words for entanglement and teleportation, we need an information media. In this model, the proposed singularity acts as such media. Further more, the above experiments defies string theory interpretation of entanglement. Because in the above experiment a particle (being string or any other shape) is not responsible for entanglement and teleportation, rather phase transitions from particle to qubits and back to particle is shown to be the actual phenomena behind the process. This is in line with the wave-particle function model proposed in Wave- Particle chapter.

Concept Review:
Quantum Fluctuation, Conservation Law and Mind

Let us discuss some of the conclusions derived from quantum theory; "In the nucleus there is a "sea of quarks", which continually pop into and out of existence in our Space-time universe due to quantum fluctuations."

This violates the conservation of energy principle that we so dearly adhere to. Here we have to look beyond the space-time and open our vision to series of related events out of our universe. Above I suggested that energy can penetrate from Planck holes into the space-time and create the sea of quarks. Also matter can transform to energy and leave our space-time through the holes. Therefore, in this view the two-way transformation of energy and matter is suggested for quantum foam phenomena.

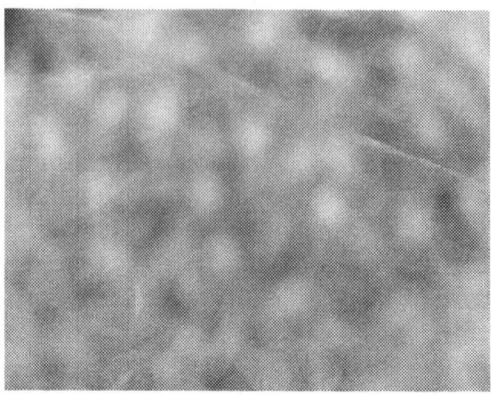

Quantum Foam

In addition, quantum fluctuation is comparable to mind activities during awakeness. During normal mind activities our consciousness is intermittently in and out of objective world. Our consciousness' constantly fluctuates between two different arenas. We are constantly alternating between sensation and imagination domains. In other words, our mind is constantly popping into and out of the physical world? We can also compare sleep dreaming and awakeness as an analogy. While we are awake we are time bonded and local. We also are dealing with objective reality most of the time. During sleep we depart from objective world and become non-local and floating out of time limits. Please refer to Quantum Mind by Arnold Mindell on the related topics.[14]

Dr. Mindell very clearly shows the many resemblances between mind activity and quantum behavior. There is no way that I can do any better.

Orbiting Electron

In the current understanding of the atom, an orbital is not something like the orbit of a planet around the sun. Orbiting electron is present at the same time in a whole region of space and has a whole range of velocities. It is a probability distribution in space. And, the electron cannot be said to be really 'moving'.

Current belief indicates that, electron does not follow a specific trajectory in its orbit to reach to a new position. This violates the locality rule in space. Experiments reveal that the electron only ever moves radially -- in and out.... But yet it covers the whole angular range. So this way, it defies steel ball rotation movement in classical physics. Alternatively, we can claim that, an electron in its orbit is everywhere and nowhere. If we cannot follow electrons trajectory, we have to come to conclusion that at times it is absent, and does not occupy a spatial location. If it does not follow the space-time rule of locality, may I suggest that electron or its trace travels to singularity and comes back to our space-time in different location according to the scenario explained in wave function chapter?

This interpretation brings the quantum mechanical experiments to more understandable and meaningful events.

Conclusion

If we take our universe as a closed system, in many quantum mechanical phenomena the first law of thermodynamics (conservation of energy) goes under question. Introducing virtual particles does not satisfy the common sense. Either we have to drop universality of conservation of energy and matter or accept another entity which interacts with our universe.

Here we have explained different quantum mechanical paradoxes based upon the proposed singularity and zero point energy derived from it. The followings are the belief of the researchers in California Institute for Physics and Astrophysics about ZPF and stochastic electrodynamics:

"In fact, two distinct views about it (Zero point energy) exist today. One justification for making such an assumption is that by adding the ZPF to classical physics many quantum phenomena can be derived without invoking the usual laws or logic of quantum mechanics. It is premature to claim that all quantum phenomena could be explained by stochastic electrodynamics (that is, classical physics plus the ZPF), but that claim may one day turn out to be the case. In that event, one would have to make a choice. One could accept the laws of classical physics as only partly true, with a wholly different set of quantum laws required to complete the laws of physics; that is essentially what is done in physics now. Or one could accept the laws of classical physics as the only necessary laws, provided they are supplemented by the presence of the ZPF."[53]

Here I adopted the second choice. I have tried to extend the classical physics law to Planck scale arena by adding singularity and the field which I postulate is derived from it.

Summary

While quantum mechanics is very precise to predict and reproduces the results of experiments, it does not offer any explanation for the strange results obtained in the lab. The proposed singularity as a non-local media, offers explanations for the strange results by defining zero, infinity, imaginary and complex numbers. Explanations for Tonomura's two slit interference, quantum entanglement and other quantum mechanical phenomena has been proposed based on this model.

Wormholes

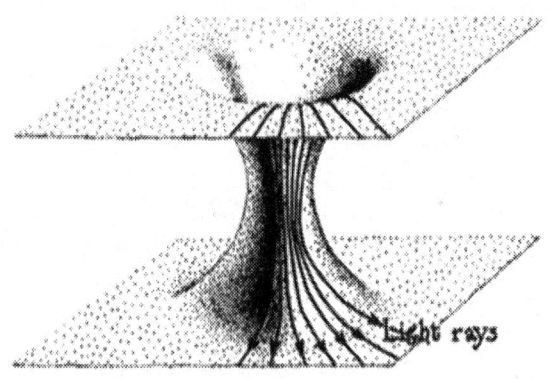

Figure 4

A wormhole is a hypothetical pipe-shape extension of space time which creates short cut between distant portions of space. Matter can pass through some of these tunnels and reaches to remote places in the universe.

Marcelo Alejandro Maidana explains the story of introduction of the wormholes and mathematical equations for them. The paragraphs and figures are copied from his interesting website. He was nice enough to give me the permission to publish these paragraphs in my book. I specifically copied it in these pages because it also demonstrates that benign singularities may be mathematically possible.

"Professor Kip Thorne, of the California Institute of Technology, was asked, in the summer of 1985 to search for alternate wormhole solutions that would allow safe passage for interstellar travelers. His motivation was a request for help from his friend and colleague, Carl Sagan. Professor Sagan was writing a science fiction story (The novel was entitled *Contact* which has been made into a motion picture staring Jodi Foster as the heroine) in which his heroine needed to cross a great interstellar distance in a very short time, namely the distance between Earth and the star Vega. Professor Thorne was only too happy to oblige.

He found a solution, which was so simple that he was surprised no one had found it before. The solution has the following metric equation.

$$ds^2 = -e^{2\Phi(r)}c^2dt^2 + \frac{dr^2}{1 - b(r)/r} + r^2(d\theta^2 + sin^2\theta d\phi^2)$$

"Where b(r) determines the spatial shape of the wormhole, and Phi(r) determines the gravitational red shift. This solution has the property of having no horizons nor excessive tidal forces to deal with which makes it safe for humans to travel through. But it does have one unfortunate drawback. In order to hold the throat open there has to be a negative energy density inside. There is no known material that has this property. Though electro-magnetic vacuum fluctuations are sometimes measured to have negative energy densities and are correspondingly called ``exotic". In order to keep the wormhole open it needs to be threaded with exotic matter that will create a tension to push the walls apart. This exotic matter would have the curious effect of defocusing light as it passed through.

"Assuming that this exotic matter can be discovered or manufactured how would one go about constructing such a wormhole? Certainly objects such as these do not occur naturally. The answer may lie in quantum mechanics. On a sufficiently small scale the universe is probabilistic. Refer to figure 5 to see possible geometries at the quantum level.

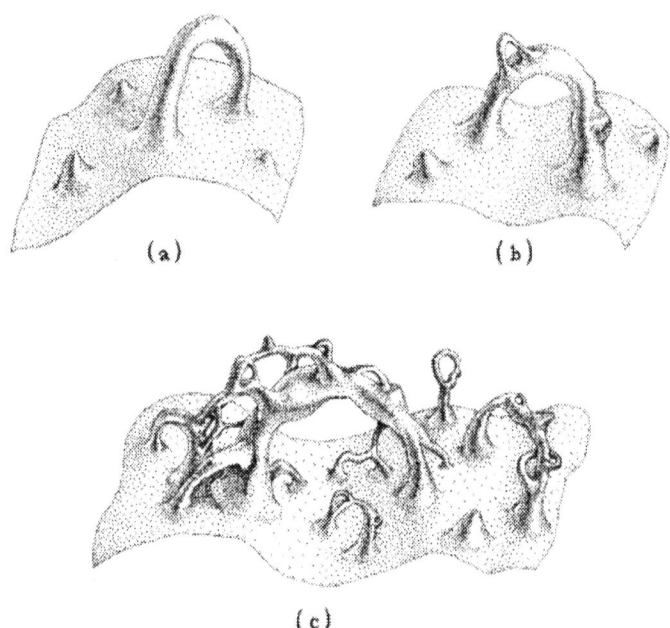

Figure 5 Embedding diagrams of the Quantum Foam
"The geometry of space-time on a Planck scale is probabilistic. The probability for (a) is 0.1%, (b) is 0.4%, and (c) is 0.02%

"There are certain probabilities that wormholes will pop in and out of existence at this level. If we assume that we, or some other society, are sufficiently advanced that we can observe this quantum foam and manufacture exotic matter, then it might be possible to reach into this microscopic universe and capture a wormhole. By pouring exotic matter into it we might be able to blow it up to a macroscopic size. We would then be poised to embark on the greatest journey imaginable."[17]

While I am eagerly waiting to see the outcome of the 1997 bet between Stephen Hawking and J.P. Preskill & Kip S. Thorne about the naked singularities,[6] I sincerely hope that Professor Thorne wins the bet so that I can claim: Professor Kip Thorne shows how singularity does not have to be bizarre. He shows that it does not have to possess horizon and can be benign.

Super String Theory

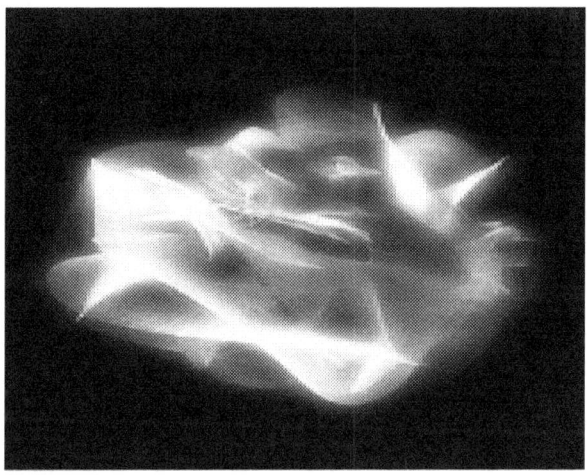

Photo / NOVA
A Calabi-Yau shape: One of the many two dimensional
visualization of the six additional spatial dimensions required by
string theory.[46]

String Theories are proposed to resolve the inconsistencies between GTR and Quantum Mechanics. Currently string theory is the prime candidate for the theory of everything. In the Standard Model of particle physics the building blocks of matter are point like zero-dimensional particles. Whereas, in string theory the building blocks are one-dimensional extended objects (strings). Strings can be open threads or closed loops. Many texts are available which explain the detail of the theory. It is beyond the scope of this book to describe the elements and principles of string theories.

One of main objectives of the super string theory is to solve the problem of the so-called space-time singularities. It tries to define a kind of world that its boundaries stop at Planck distance.

Uphill struggle has been done in order not to pass to sub Planck distances and deny zeros and infinities. The string theorists by-pass zero and singularities, by assuming that the building block of space is a string which is the size of Planck distance ($1.6 * 10^{-33}$

cm). Therefore, they suggest that there is no zero or singularity in the fabric of space. In addition, it assigns more dimensions to the space-time to be able to solve the other theoretical physics paradoxes. Steven Hawking believes:

"All one can say is whether mathematical models with extra dimensions provide a good description of the universe. We do not yet have any observation that requires extra dimension for their explanation."[6]

He does not completely reject it, he later mentions:

"String theory is good for calculating what happens when a few high energy particles collide and scatter off each other. However, they are not of much use for describing how the energy of a very large number of particles curve the universe or forms a bond state"[6]

Denying Singularity

Underneath, I will address the occasions where String Theory faces the singularity (Sub-Planck arena in this model's terms. Etienne Klein & Marc Lachieze-Rey show how string theorists tried to open the field by adding extra dimensions in order to bypass singularities to no veil. Even after adding numerous hypothetical elements to physical world, they are still facing with singularities but from a new angle.

"One of the motivations of those who are working on this theory is to get rid of troublesome singularities in field theory calculations...Quantum Theory gets around these difficulties with a method that is as artificial as it is effective. It is called renormalization...(String Theory) deals from outset with the structure of space and elementary objects...Sure enough, problems of singularities show up in a totally different way in that string theory."[2]

In a way one can claim that, Singularity is the central dilemma of contemporary physics. Brian Greene one of the main advocates of string theory in his famous book *The Elegant Universe* explains: "The whole conflict between general relativity and quantum mechanics arises from sub-Planck- Length properties of the spatial fabric"[1]

He continues: "There is a limit to how finely our conventional notion of distance can even be applied to the ultra microscopic structure of cosmos."[1]

Brian Greene rightly mentions that our conventional notion of distance cannot be applied on ultra-microscopic structure of space. If Planck length is the smallest unit of distance, anything less than that cannot be considered space. One can conclude that no notion of distance could be applied, being conventional or non-conventional, to Sub-Planck Sea.

In his recent book "The Fabric of the cosmos", Brian Greene confirms that;

> "The theory intimates that the familiar notion of space and time do not extend into the sub-Planckian realm, which suggests that space and time as we currently understand them may be mere approximations to mere fundamental concepts that still await our discovery."[69]

In my view trying to deny zero and infinity is avoiding the reality. So I have fundamental issues with string theory and because of that I am going to get very bold against it in the following paragraphs. Of course these are my personal opinions.

String Theory: A Theory of Everything

The assumed particle for curving the space-time and creating gravity is called "Graviton". The graviton has not been found experimentally but it is further assumed that it should be mass-less and should have spin-2 (have two times faster spin than photon).

In 1974 John Schwartz and Joel Schrek claimed that a mass-less spin-2 particle that is predicted by string theory is the long sought and never found graviton. They further claimed that the equations of string theory embodied a quantum mechanical description of gravity. Therefore, they declared the string theory a candidate for the theory of everything.

Remember, the theory of everything is supposed to bring the the general theory of relativity and quantum mechanics under one umbrella.

Space in String Theory

Space in the String Theory is continuous but granular. The granules being Planck distance size strings. These strings are supposedly the building block of space.

Furthermore, to respond to numerous calls from the boundaries of universe received through mathematical calculations and physical experiments, string theorists chose to introduce extra dimensions. The original assumed dimensions were wrapped and compact and ultra small, therefore they were out of our sight. String theorists are claiming that the unexplained space-time phenomena are coming from elements present in other dimensions.

Why string theorists are introducing extra dimensions? Because extra dimensions provide freedom. Imagine two-dimension figures drawn on a piece of paper. They are subject to many limitations. They cannot enjoy the third dimension freedom. That is how string theorists find relief from the limitations imposed on us by 4-dimensional space-time while trying to explain the paradoxes within space-time context. However, one dimension added by Kalusa was not enough. Therefore, string theorist had to add another 5 dimension to obtain the needed freedom to present their interpretations. In additions, in 1995 in order to explain why there are five different models of string theory, which are contradicting each other and at the same time each claimed to be the theory of everything, they introduced M-theory and took the liberty to add the 7th extra dimension.

At one time even 26 dimensions were suggested.

D-Brane and Brane-World

The complexities and difficulties of fabricating a theory of everything which is limited to familiar space-time, led the string theorist to further assumptions. They introduced kind of membranes, which are called D-branes. D-branes are hypothetical spaces, which can have up to ten dimensions with any size.

One of the most recent speculations of string theorists is the brane-world. Brane-world is a three-dimensional-brane, which embraces our universe. Here is the question. Is the brane-space continuous or discrete? If continuous then what is going to happen to lengths smaller than Planck distance and zero in such a space? Are we going through the circle of introducing new elements to eliminate zero again?

Did the string of assumptions solve the problems? Not quite.

String's Gravity denies Extra-Dimensions

In 1987, Newton introduced the inverse square law. It simply declares that the force of gravity diminishes by the square of distance between two objects.

$G = M * m / d^2$

Where M and m are masses of each object and d is the distance between them. The law simply indicates that if the distance is increased by factor of two the gravity force decreases by a factor of 4. Make the distance 3 times and gravity is decreased by a factor of 9. The inverse square law reflects the three-dimensional space. The reason being, the gravity force spreads and gets diluted in three dimensionality of space. Electromagnetic follows the same rule as well.

If we have n dimensions in space, then the gravity equation has to be written as;

$G = M * m / d^{n-1}$

In Brane-universe conjecture, while other fundamental forces are confined to three-dimensional brane-world, gravity is getting diluted in extra-dimensions as well.

The gravity decline is directly related to number of extra-dimensions. For a ten dimension space is governed by;

$G = M * m / d^{(10-1)} = M * m / d^9$

Of course, the force decline is related to the size of extra dimensions as well. Nevertheless, one would expect that the force of gravity gets diluted much greater than what inverse square law dictates. At least, such a gravity decline should reveal itself in astronomical calculations. So far, cosmological observations confirm the Newton's inverse square law. No further decline has been observed.

In 1988, the gravity law could be tested down to 1mm with the then current probes. This led Nima Arkani-Hamed, Savas Dimopolous, and Gia Devali to speculate that "in the brane-world scenario, extra dimensions could be as large as 1mm"[69]
So it was assumed that in scales less than 1mm, the inverse square law will break down. Currently, the gravity has been tested down to one tenth of the millimeter and inverse square law still holds.

Flat Universe

In addition, string theorists have assumed that the main dimensions (four dimensions in Minkowski's space) are circular in big scale (spherical universe). So they speculated that at the time of Big Bang all dimensions were just a point. As the universe expanded, space opened up and created circular dimensions. So the observable dimensions are supposedly circular. They further assumed that if this is the case, so maybe there exist other dimensions, which did not open up. These guess work might offer some solutions so string theorists could avoid zeros. However, Brian Green questions: "What if the spatial dimensions are not circular in shape? Do these remarkable conclusions about minimal spatial extent in string theory still hold? No one knows for sure."[1] However, recently cosmological constant was announced as a non-zero factor. A non-zero cosmological constant favors a flat universe,

The Cosmological Constant is a factor that was first introduced by Albert Einstein. It acts against the gravity and prevents the collapse of the universe. There is controversy about the origin and the nature of this constant. In fact, its effect surpasses the gravity and according to current believes it is creating an ever-expanding universe.

If the universe is flat, we may conclude that spatial dimensions are linear and not circular. Therefore, the base of the assumption of wrapped extra dimension is trembling.

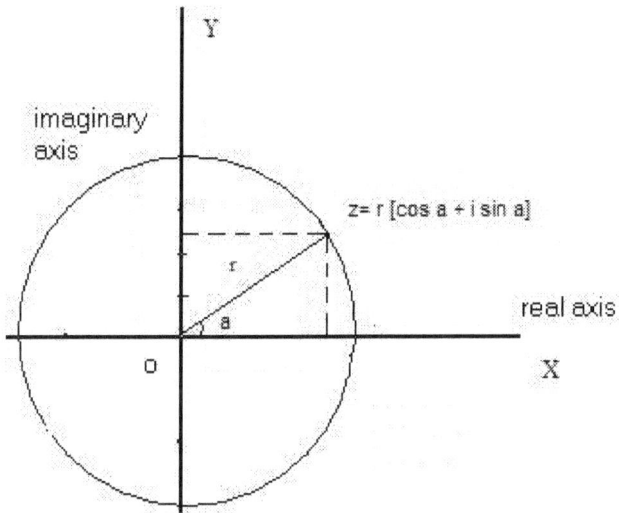

Periodic nature of Complex numbers

The notion of complex numbers indicates that any computable (matter, space and time) are fundamentally discrete (assertion C5 in this text). Assuming any extended and continuous object is against the complex number mathematics which is the basis for quantum mechanics and modern science.

Strings as Particles

Particles in string theory are one-dimensional threads which we can see only the cross section of it (as a point). Point particles in quantum mechanics on the other hand, do not posses any dimension.

The strings in the String Theory are either loops or free end. These elements with their different vibration patterns can represent different energy level. Then the energy can be translated as mass of the particle. One expects that the liberty to choose any vibration should help us to find a particular vibration which matches at least one of the known particles. The problem is, after three decades of extensive research, there has not been any resemblance between particles in string theory and the actual sub-atomic particles.

Then the next assumption came to rescue. If extra-dimensions have different sizes, then the loop's vibration in different

235

dimensions with different sizes will open a new possibility to explore. Maybe if we find the right size for each dimension, we can come up with vibrations that resemble the common particles. So the legend continues.

Then there were problems of complicated calculations so they assumed that there are symmetries in the whole elements of the universe. Calculations by using symmetry principles were easier and achievable.

The strings are so small that can not be detected so the whole idea could not be tested and therefore could not be proved. Then Super-symmetry came to picture and the assumption that there are much bigger particles which are symmetric to small and not observable strings. It created the hope that sometime in the future we can observe the bigger partners and prove the conjectures.

Brian Greene himself questions the string particles:
"Just as string theory shows that the conventional notion of zero- dimensional point particles appear to be a mathematical idealization that is not realized in real world, might it be the case that an indefinitely thin one dimensional strand is similarly a mathematical idealization."[1]

String of Assumptions
The string of assumptions does not stop here. The complexities imposed by confining ourselves to space-time arena, led the string theorists to even believe in possibility of existence of 10^{500} different worlds thus they indorsed multiverse concept.

With so many assumptions, one can find a solution for any kind of puzzle. Imagine you have the liberty to redesign a crossword puzzle by moving around the black squares to your will and select your own words to place in the squares. Moreover, you take the liberty to choose the shape and the size of the puzzle. Solving the puzzle will get very easy this way. As the problems aroused more postulations came to rescue. Now we have built a fascinating theory, which is mind wrenching and a good challenge for the boys to entertain themselves. Is this waste of talents and knowledge of brilliant physicists and mathematicians? We do not know. For solving the biggest question of century, any path has to be explored. Nevertheless, this is my question;

Can the fundamental structure of our universe be so complicated? Ray Salmonoff suggests:

"If a given set of facts about the world can be explained by more than one theory, how do we choose between them? ...The short answer is to use Occam's razor: you pick the theory with least number of independent assumptions."[7]

Maybe we'd better accept point particles as the building blocks of the universe and take violent quantum jitters as the process of two-way transformation of matter and energy at Planck's Length level (or Planks pores). Helge Kragh questions String Theory:

"Even on the theoretical level, there were several problems, namely, that theories were plagued by infinities and what are technically known as anomalies. Anomalies are terms that violate the symmetries or conservation laws when the theory is quantized. And therefore make the theory inconsistent...The entire development of super string unification was mathematical."[3]

We try to close our eyes on those mathematical results, which we dislike. We have been selective and biased. It just added to our uncertainty and confusion. Maybe the time has come to revisit the issue.

0-Dimension Singularity

Space-time settlers cannot imagine a point with no-dimension. The space-time way of thinking motivates the string theorists to assume wrapped and out of site dimensions to accommodate a sensible space-time thing. Above I suggested that we are exposed to singularity wherever there is no dimension (inside Planck Length and beyond the boundaries of universe). In twilight zone of boundaries the quantum leap is the constant communication between space-time universe and singularity. The particles (or trace of them, whatever it may mean) do not have any dimension when they are out of space-time (particle-wave chapter). In this view we do not need to assume dimensions being tangible or hidden for explaining some of the physical findings.

In 1960s Stephen Hawking and Roger Penrose introduced the Singularity Theorem which showed that a Ricci flat extra space evolved in time has to be singular. This also includes the (9+1) dimensional space-time proposed by super string theory. Interestingly the time needed for compact extra dimension to turn to singularity is at Planck time scale of 10^{-43}s. Mind you that in my model space is four-dimensional (3-space + 1-time) and is discrete at the Planck distance and time level.

In his recent book, The Road to reality, Roger Penrose writes "If we wish to...obtain a non-singular perturbation of the full (1+9)-space... then we must consider disturbances that significantly spill over into...space-time as well. But in certain respects such disturbances are even more dangerous to our ordinary picture of space-time... (Which) is in gross conflict with observation?"[56] Please note that, no-dimension provides ultimate freedom and fulfills the need to adherence to ovbservations. We just need to free ourselves of the notion of space and time and imagine a no-space-time zone.

To me the assumption of one dimensional string particle (extension of mass proper to unseen extra dimensions) and suggesting the existence of a super-space (which is still a space-like entity) beyond the familiar space-time is the frantic attempt of cliff settlers to keep their ground and not to pass the edge. Maybe we have to assume that the entity beyond the boundaries of space-time is not space-like and has no dimension. Maybe we'd better off postulate that mass beyond space time can convert to something else (energy in this model).

Summary

Although String theory is the most popular model for theory of everything, it has its shortcomings. The theory is based on denying zero and infinities. It also employs frequent independent assumptions. At the same time it builds a complex and sophisticated model for space which is hard to prove. In my view, because it tries to deny zero and infinity, it overlooks a main portion of reality.

An excellent series of videos in a layman terms presentation was produces by Brian Greene featuring Steven Weinberg and few other great physicists of our times. You can watch them at: http://www.pbs.org/wgbh/nova/elegant/program.html

Dark Matter

The Milky Way in Infrared
Credit: E. L. Wright (UCLA), The COBE Project, DIRBE, NASA

Dark matter is a kind of matter, which unlike to ordinary matter does not emit or reflect enough light, X-rays or other electromagnetic radiation. Therefore, it is not directly detectable by our instruments. However many astrophysical evidences points to the presence of such matter. In fact dark matter supposedly is much more abundant than ordinary matter. The amount of observable matter in our Milky Way galaxy is only about ten percent of the mass that is needed to keep the system stable and holds on to the stars in the outer orbits of the galaxy. Therefore, it is conjectured that %90 percent of the matter in the Milky Way galaxy is dark matter. NASA explains the origin and nature of dark matter:

"Dark matter has been a nagging problem for astronomy for more than 30 years. Stars within galaxies and galaxies within clusters move in a way that indicates there is more matter there than we can see. This unseen matter seems to be in a spherical halo that extends probably 10 times farther than the visible stellar halo around galaxies. Early proposals that the invisible matter is comprised of burnt-out stars or heavy neutrinos have not

239

panned out, and the current favorite candidates are exotic particles variously called neutrilinos, axions or other hypothetical supersymmetric particles. Because these exotic particles interact with ordinary matter through gravity only, not via electromagnetic waves, they emit no light."[38]

The concept of dark matter has been presented because the observable mass of galaxies has failed to equal the needed gravity to keep them stable. Gravity has to counteract the centrifugal force of stars in outer orbits in order to keep them stable in their place. So we assumed a kind of unobservable mass, which provides the needed extra space-wrapping.

It is difficult to figure out how concentrated localized globes of dark matter can be distributed across a galaxy and be able to establish the harmonic orbits of millions of stars contained in it. Such a matter has to be diffused in every point of galaxy to exhibit such an effect. It is the current belief that 90% of dark matter is in the shape of particles and shows at spherical halo around the galaxies.

To find the particles creating dark matter, Physicists are looking from Machos to Neutrinos.

Here is another speculation based on proposed model. Can the gravity attributed to dark matter originate from the Planck arena activities? Concentration of zero point energy can create pair of short life particles. Nevertheless, it will have enough life-time to exert the necessary gravity effect.

Accumulative gravitational effect of countless short lived particles may count for dark matter effect. Maybe we do not need to look for origin of neutrinos at the beginning of time. Maybe particles do not need to be made only at the Big Bang era. Maybe every single pore in space-time universe (Planck space) has the potential to produce mass at any time.

Please note the further we go from the center of galaxy the more disperse the stars are. This of course means less gravity from visible mass at the center of galaxy and more angular velocity and centrifugal force for the stars in periphery. At the same time more empty space to exhibit the above assumed effect.

In his recent (March 2005) article "Black holes 'do not exist'" George Chapline from Lawrence Livermore, National Laboratory in California published in Nature[38] mentioned that collapse of big stars creates a zone which differs from ordinary space-time and contains much larger vacuum energy. He calls this zone dark energy star which is different from condensed mass zero-point singularity of a black hole. Such dark energy star

"Surface corresponds to a quantum critical surface for space-time. The behavior of matter approaching such a quantum critical surface can be surmised from the behavior in the laboratory of real materials near to a quantum critical point. One prediction is that nucleons will decay upon hitting the surface of massive compact objects."[38]

This strange behavior, he says, is the signature of a 'quantum phase transition' of space-time. Chapline argues that a star doesn't simply collapse to form a black hole. He states that it is near certainty that black hole does not exist. He considers these dark energy stars the origin for expansion of the universe.

Remember that in a discrete space-time, for the expansion to materialize, the space-time units have to be built from within. Our speculations about phase transition of particles and its transformation to energy (no-mass singularity) is in line with the Chapline presentation in Texas Conference on Relativisitc Astrophysics, Stanford, California, December 12-17, 2004.

Flatness Problem

Geometry of the Universe

The shape of our universe has been matter of debate for a long-time. The main possibilities were the universe being spherical, flat or saddle like. [31]

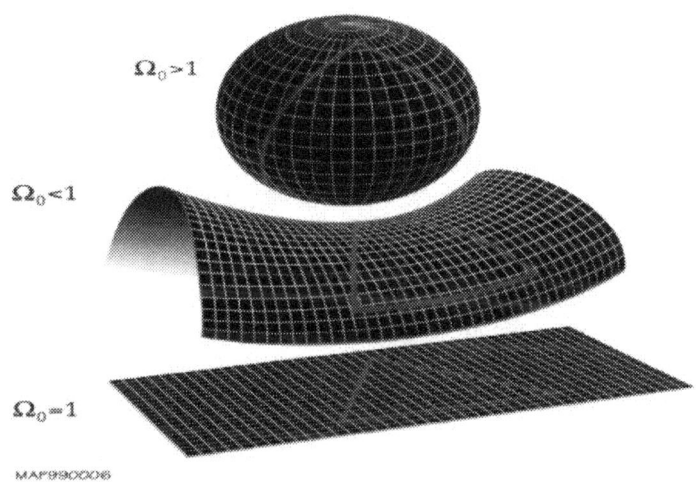

$$\Omega_0 > 1$$

$$\Omega_0 < 1$$

$$\Omega_0 = 1$$

MAP990006

Cosmological Constant

In 1917, Einstein introduced the cosmological constant in order to explain how our universe is not falling into itself because of gravity and stays static. The force presented by cosmological constant in equations supposedly counteracts the gravity forces. Therefore, it prevents the collapse and stabilizes the universe. According to its value, the constant would allow for an expanding, contracting or static universe. Since at the beginning of twentieth century universe was considered stationary, he chose a particular value for it, which would permit a static universe. The nature of this constant was not known or explained completely at the time.

When in 1929 Hubble's observation showed that universe is expanding and is not static, Einstein himself regarded the cosmological constant unnecessary, and everybody accepted a zero value for it and accepted it as a non- contributory factor.

Big Bang Theory and Expansion

The introduction of Big Bang Theory introduced the force, which is supposed to be responsible for expansion of universe despite the presence of the gravity. This would be the residual force that is left from the Big Bang explosion. It was anticipated though, that gravity would slow down and ultimately stop the expansion sometime in the future. That would be when the density of matter in the universe exceeds the critical density.

Critical density is that density of overall matter and energy in the universe that if we exceed it, the gravity has the strength to contract the universe. At critical density, the shape of universe will be flat. Above critical density, the shape of it will be convex and spherical. If the density falls bellow critical density, the shape of the universe will be concave and it will expand forever. All of this happens if we take the cosmological constant as zero. Which means if there is no other factor at work to expand the universe?

In January 1998, Alex Pilippenko in Astrophysist Conference in Southern California announced that cosmological constant could no longer be taken as zero. Different cosmologists reached to this point after long-term observation of supernovae IA in longer red shifts (far distances more than 3 billion light years away).

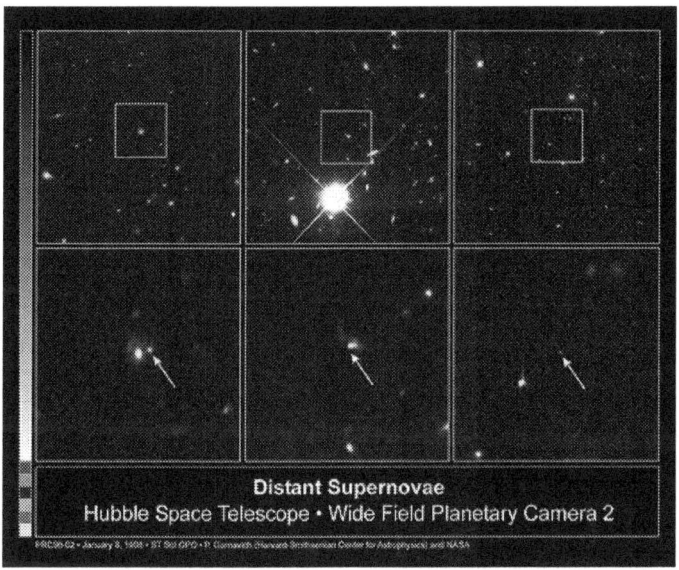

Supernovae research has been used to measure the amount of red shift and expansion rate of the universe

The Wilkinson Microwave Anisotropy Probe (WMAP) the newest probe launched by NASA in 2002 has determined, within the limits of instrument error, that the universe is almost flat. This is an explanation for the cosmologic constant effect released by NASA.

"The results of the WMAP mission and observations of distant supernova have suggested that the expansion of the universe is actually accelerating which implies the existence of a form of matter with a strong negative pressure, such as the cosmological constant. This strange form of matter is also sometimes referred to as the "dark energy". If dark energy in fact plays a significant role in the evolution of the universe, then in all likelihood the universe will continue to expand forever." [31]

Density of the Universe

Since the cosmological constant is slightly positive, we conclude that space is nearly concave. That means that Omega M (world density/critical density) has to be less than 1. Ω M <1

To have a concave universe the sum of Ω M plus the effect of cosmological constant in the expanding universe (Ω Lambda) has to be smaller than one.

Ω M + Ω Lambda <1

Here we need to assume a factor at work, which provides the positive cosmological constant.

Dark Energy
By definition, dark energy is a hypothetical force which exerts negative pressure and is acting against the gravity at large scales. This energy is not only acts against gravity and prevents Big Crunch. Supposedly, it is also responsible for accelerated expansion of the universe. We can further speculate that this entity is the source for dark matter, as well. In addition, in a discrete space-time model, this energy can also be responsible for creation of space-time blocks.

Dark Energy in this Model
Previously, I have claimed that the source of dark energy is inherent in this model. In Singularity chapter, I have postulated that Zero Point Energy is out of our space-time and is an element of proposed singularity. An out of space ZPF can open our eye to new possibilities and scenarios.

This scenario gives us the good news that big crunch will never happen, because the universe is going to expand forever. Then there comes this feeling of emptiness and the chilling sensation that our descendants will stay lonely in an infinitely vast and cold universe. One may say that, our own galaxy or branch of galaxies, which remain together, has many new places to explore. Why do we have to be so ambitious and greedy to dream about intergalactic traveling? Is there an end to human race greed? Is there an end to evolution's ambition? Are we a dead end branch of evolution? Is there an alternative for cold and disperse universe scenario in the future?

Zero Point Energy
Ever expanding universe also implies that in a universe with a discrete fabric, space-time units constantly have o be created. In

addition, to keep the matter density near critical value, matter has to constantly appear in the newly formed space. For the above to happen, ZPE has to constantly penetrate inside the universe, create space-time blocks, and matter particles. Energy permeation from every miniscule of space forces universe to exponentially expand. Can there be any other explanation for the acceleration? According to Donald Goldsmith, no:

> "Unless we are prepared to reject the tenets of general relativity theory, the only additional term that may appear to produce an acceleration consists of...a cosmological constant of unknown size, but one with the properties we have described – namely, a transparent energy, untappable and untouchable, so far as we can tell, except by its tendency to make the universe expand more rapidly."[29]

In 1990, Hideo Kodama of Kyoto University demonstrated the positive energy density equations. It was based on assumption that cosmological constant is positive. In this case, it:

> "predicts a spectrum of quantum fluctuations in space-time. It also has a precise Planck-scale description, which makes use of very elegant mathematics connected to the invariants of graphs and knots."[48]

Alfonso Rueda from department of Electrical Engineering California State University suggests that:

> "Both mass and the wave nature of matter can be traced back to specific interactions with the electromagnetic zero-point field and possibly the other bosonic vacuum fields. Given this possible reinterpretation of fundamental properties, we suggest that it is premature to take a firm stand against the reality of the zero-point radiation field and its associated energy on the basis of cosmological arguments, especially given the possible relation between quantum vacuum, or zero-point, radiation and dark energy."[55]

The Conservation Energy Law

Nonzero cosmological constant require infinite source of energy readily available in every miniscule of space. Where does this energy come from? Whatever happened to conservation of energy law?

The first law of thermodynamics (the conservation of energy) states that the total inflow of energy and matter into a closed system must equal the total outflow of energy and matter from the system. It predicts that energy can be converted from one form to another, but it cannot be created or destroyed. If we take the space-time universe as a closed system, we are faced with a big puzzle. How are we going to offer a solution for expanding force, if there is no possibility for creation of energy inside the universe?

Some physicist try to disprove the presence of dark energy by suggesting models like modified gravity as proposed by Damien Easson or even dark geometry as speculated by by Gu-Je-An presented in Cosmo 4 in Toronto, in 2004. Others try to demystify its source and nature. In the proposed model, the so called dark energy is inherently built in into the scenario and because we took the universe as an open system, its source is explained.

Moreover in line with $E = mc^2$ this energy upon changing to matter, has also the potential to positively contributes to curvature of space (dark matter). As mentioned above, the nature of dark matter is not known. No body knows if it is similar to ordinary matter or it is antimatter or made of super symmetric particles. If dark matter is ordinary matter it can contribute to star formation.

The Lagoon Nebula consists of hot star dusts which by contraction create new stars

Steady State Theory

The Steady State theory is a model developed in 1948 by Fred Hoyle, Thomas Gold, Hermann Bondi. In the Steady State model, new matter is continuously created as the universe expands. Therefore, universe can remain in steady and stable condition. After discovery of microwave background radiation, astrophysicists gradually distanced themselves from the theory in favor of the Big Bang model for creation of universe.

The Hubble Deep Field photograph taken in 1996 by the Hubble Space Telescope shows the most distant view known so far. It was expected to show the birth of galaxies, but instead shows galaxies looking remarkably like the present ones. This is in favor of steady state theory. The theory claims that universe has been steady during its history. Although the concept has been discredited by numerous evidences, perhaps a modification of that model can still be useful. Narlikar, Hoyle, Barbidge and some other astrophysicist introduced the Quasi Steady state theory to offer solutions for energy and matter creation through many small

249

Big Bang events, which has been happening throughout the age of universe. Narlikar, Hoyle and Barbidge claim;

"Creation of matter is governed by a conservation law which operates to prevent space–time singularities, which otherwise occur in general relativity...Unless creation of matter is included in physical laws, the laws lack universality."[34]

The amount required to preserve the steady state is undetectably small—about a few atoms for every cubic mile each year or roughly a few hundred atoms of hydrogen in the Milky Way Galaxy each year. The required dark energy to maintain the present expansion of universe also is very small.

Trying to deny the acceleration of universe expansion by steady state theorists proves to be a very difficult task. Acceleration has been shown by different methods and recently by WMAP observations. May be the theory have to be modified to accommodate the new observations. This may mean radical and fundamental change. Maybe we need a combination of modified steady state and the Big Bang scenario?

Nancy Kerrigan Problem
So far, supernovae observation revealed that:

Ω M + Ω Lambda = approximate to one
Ω Lambda – Ω M = approximately 0.4

The above figures are obtained by different methods like Supernova Observation, Cosmic Background Radiation studies, Gravitational Lensing and so on. Although the above numbers are not firm yet and still under investigation, they are the cosmologist's best guess.

According to the Big Bang Theory, at the beginning, the matter density was much higher and cosmological constant was much smaller. Gradually matter got less dense and cosmological constant increased. We are at the point where Omega M and Omega Lambda are almost equal. This creates a puzzling issue. Out of all the possible combinations, why should we live in such a period which is favorable to life? This is what Robert Kirshner called Nancy Kerrigan problem referring to famous figure skater after the attack instigated by skating rival, "Why me? Why now?"

Roger Penrose in his resent book *The Road to Reality* a writes: "The seeming coincidence that Ω lambda and Ω M are of the same general order of size seems like a puzzling coincidence."[56]

We can either hang on to philosophical answers like weak or strong Anthropic principles, which claim human only can live in this period. In any other periods, the conditions were not suitable for a human to be alive and ask these questions. Or explore the possibility that a stable relation between matter density and cosmological constant effect has existed from the beginning to date and will continue in the future. This can happen if energy has been steadily penetrating inside empty space.

If we take the size of space after initial expansion, as variable and matter density as constant then the energy penetration through more space pores in expanding space will offer an explanation for Ω lambda and Ω M to be at the same general order. So one can speculate that may be after initial expansion the matter density remained the same by dark energy penetration which compensated the space expansion.

Summary

Here we adopted the zero point fields as the source of dark energy. As mentioned before, the objection against the notion of zero point field is rooted to the fact that such a field is expected to have interaction with electromagnetic radiation inside. This effect is not observed. In this model zero point energy is out of space-time so it bypasses the above problem.

Previously, the question raised that; is there an alternative for cold and disperse universe scenario in the future? If we have constant energy penetration and matter formation in the universe, the future of world may not be as disappointing after all.

Strong and Weak Anthropic Principles and Multiverse Model[63]

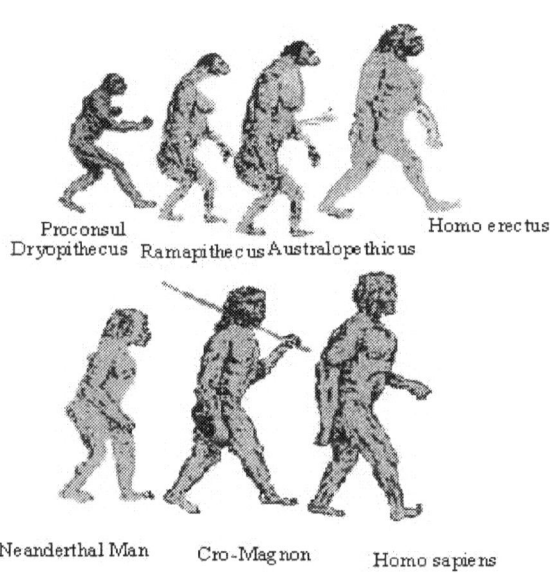

Proconsul   Homo erectus
Dryopithecus Ramapithecus Australopethicus

Neanderthal Man   Cro-Magnon   Homo sapiens

http://www.pinkmonkey.com

The world surrounding us is organized, coherent and sophisticated. It also has evolved in a way that life can appear in it and humans can roam at least the planet earth. The history of the universe also shows that at the side, a path to more organization and sophistication has been followed. From elementary particles to hydrogen and helium atoms and heavier elements, from simple inorganic molecules to complex organic ones, the history of our world is full of wonders. These wonders create lots of questions. Why out of all of the possibilities the world followed such a sophisticated path and why it should be so compassionately hospitable to human beings? Why?

One of the possible answers is called *Weak Anthropic Principle*. It states that for us being here, all of the complexity established during the history of universe had to take place. If it didn't we wouldn't be here. Because it happened as it happened, we are here and can ask the questions we ask. In any other kind of history we could not exist today to ask. So although such a history has been very improbable and remote, it happened and we are here today.

The other answer is called *Strong Anthropic Principle*. It states that in order for the cosmos to be as it is, somehow it requires the influence of an observer. Religion advocates normally use this kind of interpretation. But there are scientific extrapolations using this principle as well. One of the views speculates that the macro world that we are observing is that portion of reality that we can see, measure and have the knowledge about. The actual reality is deeper and exists in quantum level in superposition of states. We, according to our state of mind can see a portion, which suits us. So in a way we may say that without us the universe does not exist as we see it.

The other possible solution is what Martin Rees names it multiverse option. It proposes multi-universes, which can melt and transform to each other and we are living in one of them, which is hospitable to life. This is almost impossible to prove or disapprove. This solution seems more imaginary than a scientific theory.

None of the above answers offers a strong argument. In my opinion, the problem is, we confine ourselves to space-time and limit the physical laws to it. Because of this, many problems remain unsolved or we have to travel far to the outskirts of logic, in order to offer an explanation for them.

It is understandable if scientists try to avoid theological approach to reality. That kind of approach is not scientific or logical. However, avoiding a big portion of reality, just because we are afraid that it may be interpreted as presence of God, is not acceptable either. I do not understand how developing theories, which include out of our space-time elements, can disprove science. Are we scared of unknown just like fifteenth century priests? Are we hiding in our caves? I believe, understanding the laws of entity beyond our space-time are attainable and we do not have to refer to God of the Gaps to find a solution for our questions.

We have enough scientific means to explore the out of universe entity and build theories in the basis of dual nature of existence. Maybe we need to look for the answers in such a dual setting.

Conclusion

General Relativity vs. Quantum Mechanics
I have claimed that this is a model for theory of everything. But so far I have not offered a direct answer for the central question.

Why is there incompatibility between the two pillars of physics? This question has created much controversy over the past century. In specific terms, if gravity dictates a smooth curved spatial geometry (what we actually encounter in macrocosm), why is it that in Planck scale we face such severe distortions and wrapping of space-time and quantum frenzy?

This view offers the following answer;

We postulate that our space-time universe is exposed to a mass-less singularity in every Planck hole. In other words, Planck scale is the boundary of the space-time universe in micro scale. In such a paradigm, gravity is not the only factor. The fabric of universe in Planck scale is severely affected by energy from singularity. This energy demonstrates deflective force. Since according to Grand Unified theory, in Planck scale different forces are supposed to be unified, this energy is invariant and different forces are in a way in superposition of states. We further have speculated that particles in their wave motion are under influence of this force. Since in majority of occasions the particles have to return the energy that they borrowed, back to singularity during their wave function, the turbulence does not extend far beyond Planck Scale, therefore we have smooth and curved spatial geometry in large scale.

We further speculated that the elasticity of the fabric of space is the Counter-force, which returns the particles back to singularity.

Zeros and Infinities
One would ask why there is the tendency to avoid zeros and infinities in theoretical physics. The answer is these domains cannot be designated to any entity inside space-time universe. Our world is finite and zero does not point to any quantity so it cannot be defined in space-time. But the fact is they are very powerful

elements in our mathematical calculations. So we have to embrace them. If we cannot accommodate them inside our universe may be we have to accept their existence outside it. Modern physics and specially Quantum Mechanics in addition to our everyday experience suggest that there are factors affecting our everyday endeavor, which we cannot attribute them to known physics. Just like zero and infinity which are affecting math calculations but we cannot assign them to any element in space-time. Therefore, we have to open our scopes to new horizons.

On the other hand, many believe that if an out of universe entity exists we cannot observe or understand it. So they prefer to leave it untouched or ignore it. The presented model in particle wave chapter suggests that we are constantly in contact with this entity in more profound level. So by this definition, singularity should be comprehensible.

Others believe that if we accept a kind of outside entity affecting our world, we have to blindly accept a supreme power that we don't have any clue about it. This would imply that science accepts religion.

Lee Smolin describes it very clearly. He denotes, if we accept that amongst countless possibilities, our universe chose the one, which is hospitable to life, we have to accept that something outside our universe made the decision:

"This is the exact point at which science becomes religion. Or to put it better, it will be rational to use science as an argument for religion."[27]

Of course, linking to religion with all of its shortcomings is not acceptable to mainstream science. Thus, this choice is out of question.

That reminds me how fifteenth century priests were so afraid of any new scientific thought or finding. They were afraid that it would deny religion and god. So they religiously fought against the new scientific findings to the point that they would kill the scientific belief and the believer at the same time, if it was needed. They were scared and did not want to come out of their cave. One wonders, what would happen to their god if Galileo showed that the earth is round?

Do we avoid, kill and normalize a big portion of our findings and evidences just because it may shake our philosophical belief?

Humans are truth hunters. Slowly but surely we will get access to deeper levels of reality.

As I mentioned, one reason for objections that is normally put forward is that, anything out of our universe is not accessible to us. Thus, we cannot observe and understand it or experiment with it. So it does not fit inside scientific domain.

Above, I tried to show that some of the characteristic of the entity outside of the space-time is debatable by scientific methods. Quantum mechanics, positive cosmological constant, dark energy, dark matter, black holes and transpersonal psychology are expanding our horizon to the boundaries and beyond. Is it time to expand our exploration beyond the boundaries? Isn't it the time to at least fantasize, and hypothesize theories, which expand beyond space-time? I believe, if it offers logical solutions to our questions, it worth the speculation.

Last Word

A new way of looking at the physical world is presented in this view. Roger Penrose suggests that:

"Physics has to find an objective reduction theory to connect Quantum level to Classical level. While these two levels are computable, the new theory has to be non-computable and non-local."[5]

It seems this view meets the above requirements. Because the presented concept has the potential to offer solutions to our long lasting questions, it deserves further attention and speculation. This view not only can offer solutions for a number of unanswered problems in physics, but it has the potential to shed light onto many mysteries such as life and death, mind, life sciences and psychology, Para-psychological findings, etc.

The concepts presented may not only change the way that we look at our world, it also can change the way we live.

References

1) Greene, Brian R. *The Elegant Universe.* Vintage Books, 2000.
2) Klein, Etienne and Marc Lachieze-Rey. *The Quest for Unity.* Oxford University Press, 1999.
3) Kragh, Helge. *Quantum Generations.* Princeton University Press, 1999.
4) Earman, John. *Bangs, Crunches, Whimpers, and Shrieks.* Oxford University Press, 1995.
5) Penrose, Roger, et al. *The Large, the Small and the Human Mind.* Cambridge University Press, 1955.
6) Hawking, Stephen. *The Universe in a Nutshell.* Bantam Books, 2001.
7) Davies, Paul. *The Mind of God.* Touchstone Books, Simon & Schuster, 1992
8) Walker, Evan Harris. *The Physics of Consciousness.* Perseus Publishing, 2000.
9) Clarke, Chris. *The No-Locality of Mind.* Tues. Feb. 4 16:22:05 GMT, 1997.
10) Stenger, Victor J. *Physics and Psychics.* Prometheus Books, 1990.
11) Greenstein, George and Arthur G. Zajong. *The Quantum Challenge.* Jones and Bartlett Publishers, 2001.
12) Levy, David H. *Book of the Cosmos.* St. Martin's Press, 2000.
13) Hogan, Craig J. and Springer Verlag *The Little Book of Big Bang.* New York Inc., 1998.
14) Mindell, Arnold. *Quantum Mind.* Lao Tse Press, 2000.
15) www.nature.com/nsu/031013/031013-10.html
16) Andrew Hamilton's home page. Center of astrophysics and space astronomy. University of Colorado. http://casa.colorado.edu/~ajsh/schww.html
17) http://www.geocities.com/CapeCanaveral/Hall/5803/tra.html
18) http://www.damtp.cam.ac.uk/user/gr/public/holo/
19) Prideaux, Jeff. ACSA and the BCN GROUP. http://www.acsa.net/
20) http://www.merkabaweb.net/holo2.htm
21) http://users.erols.com/iri/ZPENERGY.html
22) http://users.erols.com/iri/ZPENERGY.html
23) Dr. Puthoff, H. E. Institute for Advanced Studies. Austin, Texas. http://www.ldolphin.org/zpe.html
24) http://www.wolframscience.com/nksonline/page-1060a-text
25) *Nature Journal.* 11 March, 2004. Page141-144.
26) http://roxanne.roxanne.org/epr/index.html - table

Mohsen Kermanshahi

27) http://www.glenbrook.k12.il.us/gbssci/phys/Class/refrn/u14l5da.html

28) Smolin, Lee. *Quantum Gravity*. Basic Books, Perseus Books Group, 2001.

29) http://gregegan.customer.netspace.net.au/SCHILD/Connect/Connect.html

30) Goldsmith, Donald. *The Runaway Universe*. Perseus Books, 2000.

31) http://map.gsfc.nasa.gov/m_mm/mr_content.html

32) http://map.gsfc.nasa.gov/m_uni/uni_101shape.html

33) http://www.schoolsobservatory.org.uk/study/sci/cosmo/internal/stea dy.htm

34) http://en.wikipedia.org/wiki/Conservation_of_energy

35) http://www.geocities.com/CapeCanaveral/Launchpad/8098/Hoyle.htm

36) http://plato.stanford.edu/archives/win2002/entries/qm-bohm

37) http://news.nationalgeographic.com/news/2004/08/0818_040818_tel eportation.html

38) http://universe.nasa.gov/press/2003/031105a.html

39) http://hubblesite.org/newscenter/newsdesk/archive/releases/2004/12/

40) Gozin, N. Sahel. Alzahra University. Tehran, Iran.

41) http://www.crownedanarchist.com/relativity.htm

42) http://www.geocities.com/bigshrink2000/

43) http://axion.physics.ubc.ca/rebel.html

44) http://content.karger.com/ProdukteDB/produkte.asp?Aktion=ShowP DF&ProduktNr=224242&Ausgabe=230415&ArtikelNr=80557&file name=80557.pdf

45) http://www.intuition.org/txt/pribram.htm

46) http://web.mit.edu/newsoffice/tt/2003/oct22/elegant.html

47) Smolin, Lee. *Scientific American*. January, 2004.

48) http://board.dserver.org/n/nuphys/00000011.html

49) Weinberg, Steven. "A unified physics by 2050?" *Scientific American*. Volume 13, Number 1.

50) Kane, Gordon. *Super Symmetry*. Perseus Publishing, 2000.

51) California Institute for Physics and Astrophysics.

52) Feynman, Richard P. *Quantum Electrodynamics (QED)*. Princeton University Press, 1985.

53) http://www.calphysics.org/mass.html

54) Haisch, Bernard, et al. Update on an Electromagnetic Basis for Inertia, Gravitation, the Principle of Equivalence, Spin and Particle Mass Ratios in Amer. Inst. Physics Conf. Proc., Space Technology and Applications International Forum (STAIF-2003), Ed. Mohamed S. El-Genk, pp. 922 - 931, gr-qc/0209016 (2003).

55) Haisch, Bernard and Alfonso Rueda. "Gravity and the Quantum Vacuum Inertia Hypothesis." *Annalen der Physik.* Vol. 14, No. 8, 479-498 (2005).
56) Penrose, Roger. *The Road to Reality.* Jonathan Cape, London, 2004.
57) Pribram, Karl. http://gestalttheory.net/conv/prib.html
58) Talbot, Michael. http://twm.co.nz/hologram.html#Karl%20Pribram
59) http://www.big-bang-theory.com
60) McCutcheon, Mark. *The Final Theory.* Universal Publishers, 2004.
61) Grof, Stanislav. *Psychology of the Future.* State University of New York Press, 2000.
62) Sadra, Mulla. (Sadr al-Din Muhammad al-Shirazi) (1571/2-1640).
63) http://www.muslimphilosophy.com/ip/rep/H027.htm
64) http://www.pinkmonkey.com
65) /http://www.phy.uct.ac.za/courses/phy400w/particle/higgs3.htm
66) CIPA http://www.calphysics.org/mass.html
67) Wesson, Paul S. "Zero-point fields, Gravitation and new physics." University of Waterloo, Canada.
    http://www.calphysics.org/articles/wesson.pdf
68) http://board.dserver.org/n/nuphys/00000011.html
69) Greene, Brian. *The fabric of the Cosmos.* Vintage Books. 2004

Index

Singularity, 5, 6, 15, 33, 34, 37, 38, 39, 42, 44, 46, 50, 51, 54, 64, 79, 80, 95, 159, 165, 204, 209, 213, 217, 220, 230, 237, 246

Space, 5, 37, 42, 44, 65, 79, 80, 87, 95, 128, 129, 130, 134, 147, 185, 220, 231, 237, 249, 262

Space-Time Universe, 79

Spin Network, 91

Stochastic Electrodynamics, 157, 165

Strong Anthropic Principle, 254

Tachyon, 5, 157

Time, 5, 44, 52, 65, 80, 89, 128, 131, 134, 148

Tonomura, 6, 131, 206, 207, 209, 223

Transpersonal psychology, 61, 63

Universe, 5, 88, 108, 120, 121, 131, 163, 230, 243, 245, 249, 261, 262

Weak Anthropic Principle, 253

Zero, 5, 6, 31, 39, 46, 48, 81, 134, 156, 165, 168, 222, 246, 247, 263

Zero Point Energy, 6, 156, 165, 247

ZPE, 6, 48, 50, 156, 165, 166, 167, 171, 176, 177, 187

www.ingramcontent.com/pod-product-compliance
Lightning Source LLC
Chambersburg PA
CBHW071406170526
45165CB00001B/189